IFMBE Proceedings

Volume 48

The International Federation for Medical and Biological Engineering, IFMBE, is a federation of national and transnational organizations representing internationally the interests of medical and biological engineering and sciences. The IFMBE is a non-profit organization fostering the creation, dissemination and application of medical and biological engineering knowledge and the management of technology for improved health and quality of life. Its activities include participation in the formulation of public policy and the dissemination of information through publications and forums. Within the field of medical, clinical, and biological engineering, IFMBE's aims are to encourage research and the application of knowledge, and to disseminate information and promote collaboration. The objectives of the IFMBE are scientific, technological, literary, and educational.

The IFMBE is a WHO accredited NGO covering the full range of biomedical and clinical engineering, healthcare, healthcare technology and management. It is representing through its 60 member societies some 120.000 professionals involved in the various issues of improved health and health care delivery.

More information about this series at http://www.springer.com/series/7403

Henrik Mindedal · Mikael Persson
Editors

16th Nordic-Baltic Conference on Biomedical Engineering

16. NBC & 10. MTD 2014 Joint Conferences, October 14–16, 2014, Gothenburg, Sweden

Springer

Editors
Henrik Mindedal
Sahlgrenska University Hospital
Gothenburg
Sweden

Mikael Persson
Department of Signals and Systems
Chalmers University of Technology
Gothenburg
Sweden

ISSN 1680-0737 ISSN 1433-9277 (electronic)
ISBN 978-3-319-12966-2 ISBN 978-3-319-12967-9 (eBook)
DOI 10.1007/978-3-319-12967-9

Library of Congress Control Number: 2014953109

Springer Cham Heidelberg New York Dordrecht London

The IFMBE Proceedings is an Official Publication of the International Federation for Medical and Biological Engineering (IFMBE)

Printed on acid-free paper

Springer is part of Springer Science+Business Media (www.springer.com)

Table of Contents

Evaluation of Wavelet Based Methods in Removing Motion Artifact from ECG Signal

S. Abbaspour[1], H. Gholamhosseini[2], and M. Linden[1]

[1] School of Innovation, Design and Engineering, Mälardalen University, Västerås, Sweden
[2] School of Engineering, Auckland University of Technology, Auckland, New Zealand

Abstract— **Accurate recording and precise analysis of the electrocardiogram (ECG) signals are crucial in the pathophysiological study and clinical treatment. These recordings are often corrupted by different artifacts. The aim of this study is to propose two different methods, wavelet transform based on nonlinear thresholding and a combination method using wavelet and independent component analysis (ICA), to remove motion artifact from ECG signals. To evaluate the performance of the proposed methods, the developed techniques are applied to the real and simulated ECG data. The results of this evaluation are presented using quantitative and qualitative criteria. The results show that the proposed methods are able to reduce motion artifacts in ECG signals. Signal to noise ratio (SNR) of the wavelet technique is equal to 13.85. The wavelet-ICA method performed better with SNR of 14.23.**

Keywords—**ECG, Motion artifact, Wavelet transform, ICA.**

I. Introduction

Among various biological signals, electrocardiography is important for diagnosis of heart disease. However, it is not always possible to record clean ECG signals and a simple movement by the patient creates a significant artifact [1]. The most common sources of noise include baseline wander, power-line interference, motion artifact and electromyogram (EMG) noise. Among these artifacts, motion artifact is the most difficult noise to remove [2] because the spectrum of motion artifact completely overlaps with the ECG signal when the subject is walking or running.

Several studies proposed different methods to remove motion artifacts from ECG signals [1-4]. These methods include multi-resolution thresholding [1], wavelet based methods [2], adaptive recurrent filter [2], high-pass filter (HPF) [3], cubic spline filter [3], adaptive filter [4], independent component analysis (ICA) and wiener filter [4].

Adaptive filter is an automatic artifact removal method that operates on-line without preprocessing or calibration [4]. However, the requirement of additional sensor to provide reference input is one of the drawbacks of this method because it can add to the complexity of the hardware system [4]. Also, adaptive filter is not suitable for clinical application due to the heavy computation burden [5]. Although the adaptive recurrent filtering is able to reduce noise from signals that have repetitively morphology, but some ECG signals are not equally repeated [2]. The ICA method can operate in the non-linear domain, but it requires multi-channel signals [4]. In artifact removal using polynomial fitting, QRS complexes first should be detected [3]. Therefore, an accurate method is needed to detect QRS complexes. Wiener filter is an automatic method that can operate on single channels, but this filter cannot be used on-line [4].

The main goal of this work is to propose efficient methods based on wavelet transform to remove motion artifacts from ECG signals and to overcome the shortcomings of the previous methods. Hence, wavelet transform based on nonlinear thresholding and automated wavelet-ICA methods have been proposed. To evaluate these techniques, firstly, ECG data with motion artifact are collected from MIT-BIH database. Secondly, the proposed methods are applied to these signals to remove motion artifact. After applying each method, a low-pass filter (LPF) is used to remove high frequency artifacts from ECG signals. Finally, the performance of the proposed methods is evaluated with quantitative and qualitative parameters including signal to noise ratio, relative error (RE), cross correlation (CC) and power spectrum density (PSD) [5].

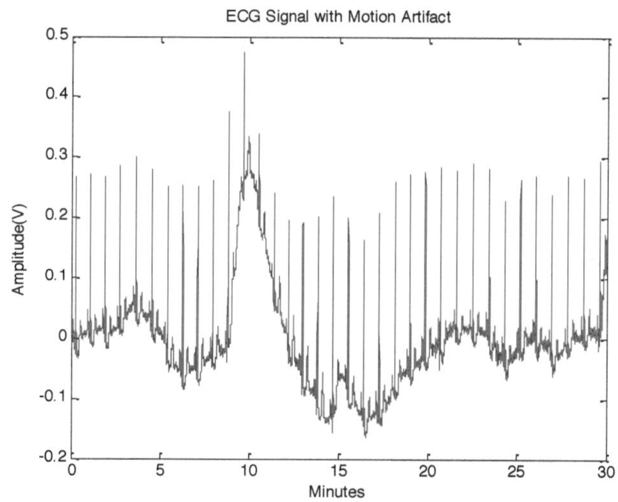

Fig. 1 ECG signal with motion artifact

H. Mindedal and M. Persson (eds.), *16th Nordic-Baltic Conference on Biomedical Engineering*,
IFMBE Proceedings 48, DOI: 10.1007/978-3-319-12967-9_1

II. MATERIAL AND METHODS

A. ECG signal collecting

The electrocardiogram signals were collected from MIT-BIH database (number 101 and 103 from healthy subjects with a sampling frequency of 360 Hz). 30 minutes of the ECG signals that include motion artifact were selected, see Figure 1. In these signals, the direct current (DC) value first was removed. To evaluate the performance of the proposed methods with quantitative and qualitative criteria it was required to simulate distorted ECG signal. Hence, the motion artifact signal was first extracted from a noisy ECG signal where both noise and clean components were well defined. Then the extracted motion artifact was added to 30 minutes of another clean ECG signal to simulate the distorted ECG signal.

B. Wavelet transform based on nonlinear thresholding

Wavelet transform is a time-frequency technique that was introduced to overcome the limitations in time and frequency resolution [6]. Wavelet decomposes a signal into different frequency components and studies each component with a resolution matched to its scale [7]. This property can be used for denoising purposes. The continuous wavelet transform formula is [5]:

$$\tilde{f}(a,b) = \int_{-\infty}^{+\infty} \bar{\varphi}_{a,b}(t) f(t) dt \qquad (1)$$

Where $\varphi(t)$ is a basic waveform that can be modified to basic functions, $\varphi_{a,b}(t)$, obtained from dilations and shifts of the basic waveform [5] and $\bar{\varphi}_{a,b}(t)$ is complex conjugate of the basic function. In equation (2), $\varphi_{a,b}(t)$ is defined as [5].

$$\varphi_{a,b}(t) = \varphi_a(t-b) = |a|^{-\frac{1}{2}} \varphi(\tfrac{t-b}{a}) \qquad (2)$$

In this study, a discrete wavelet transform using the eight-order symlet wavelet was applied to a single channel recording to create an eight-level wavelet decomposition of the raw signal, see Figure 2. Choosing this wavelet was due to similarity of the wavelet function to the ECG signal that it helps to remove noise successfully [2]. Equation (3) shows the symlet wavelet family [7].

$$h(z) = \sum_k h_k z^{-k}, \; g(z) = zh(-z^{-1}) \qquad (3)$$

Which h and g are the wavelet decomposition filters, h is a low-pass filter and g is a high-pass filter [7]. After calculating the wavelet coefficients, nonlinear thresholding in the wavelet domain was used for artifact removal, where the absolute value of the coefficients greater than the threshold was set to zero. The inverse wavelet transform was then implemented using the new coefficients to obtain the clean ECG signal [5]. Figure 2 shows four levels (four rows) of wavelet decompositions before and after thresholding.

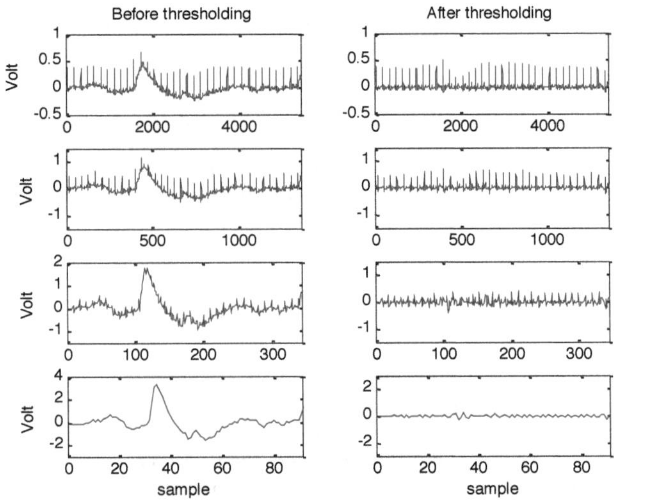

Fig. 2 Four levels of wavelet decompositions of the ECG signal before and after thresholding

C. Automated wavelet-ICA technique

Wavelet-ICA combines two techniques, wavelet transform and ICA, to overcome shortcomings of each technique. Independent component analysis extracts statistically independent components from a set of measured signals [6]. One of the limitations of this technique is the need to have multi-channel signals. Since, the wavelet creates multidimensional data from a single-channel measurement; wavelet-ICA can be used for a single-channel signal [4]. To use ICA method, the following model was considered:

$$x = As \qquad (4)$$

Where x is a linear mixture, S is independent source signal and A is full rank mixing matrix. As there are different source, x, A and s are matrixes as follow [8].

$$x = \sum_{i=1}^{n} a_i s_i \qquad (5)$$

The ICA model describes how the observed data are generated by a process of mixing the components [8]. The goal of ICA is to estimate the mixing matrix and/or the independent source signal s from the measured data x. ICA makes the assumption that different sources are statistically independent and uses this assumption to estimate the mixing matrix recursively [6].

To remove motion artifact using this method, first, discrete wavelet transform was applied to a single channel ECG signal with motion artifact to create wavelet decompositions of the raw signal. Second, wavelet decompositions of the raw signal were used as the input to the fast ICA algorithm. In the next step, to determine the component includes motion artifact automatically, each ICA component (with 1362 samples) was segmented into non-overlap frames of 227 samples. The variances of these frames for each ICA component were calculated. Also, the variance of these variance values was obtained. The component with maximum variance was considered as a noisy component [9]. To remove artifact from this component, a HPF with cutoff frequency of 5 Hz was used. Finally, inverse wavelet transform was applied to reconstruct cleaned ECG signal. After applying each wavelet and wavelet-ICA methods, a LPF with cutoff frequency of 50 Hz was used to remove high frequency artifacts from ECG signals.

III. RESULTS

In this paper, first, the wavelet transform has been used with nonlinear thresholding in order to eliminate motion artifact from ECG signal. In this method, ECG was first decomposed into eight levels by eight-order symlet wavelet to sequences representing different frequency components of the signal. The wavelet coefficients in the low frequency scales then underwent in a nonlinear thresholding process. In this step, a predetermined threshold was considered which the absolute value of coefficients greater than the threshold was set to zero. The inverse wavelet transform was then implemented using the new coefficients to obtain the cleaned ECG signal. Subsequently, LPF with cutoff frequency 50 Hz (Finite Impulse Response (FIR), Hamming window, order=100) was used to remove high frequency noise in ECG signal.

In the wavelet-ICA method, first, ECG signal was decomposed using wavelet technique to produce multi-channels ECG signal for the ICA method. After decomposing the signal in eight level using symlet wavelet, the fast ICA algorithm was used to separate desired signal from artifact. In this step a statistical method based on variance was used to find the noisy ICA component automatically. Since, motion artifact is a low frequency noise it is possible to remove it from the noisy ICA component using a HPF with cutoff frequency of 5 Hz (FIR, Hamming window, order: 100). In the next step, inverse wavelet transform was used to obtain the clean signal. Finally, LPF with cutoff frequency of 50 Hz was applied to remove high frequency artifacts.

SNR, RE and CC of the wavelet transform and wavelet-ICA for record numbers 101 and 103 were calculated to evaluate the performance of the proposed methods (Table 1).

Table 1 SNR, RE and CC for wavelet and wavelet-ICA methods for data number 101 and 103

	SNR		RE		CC	
Methods	101	103	101	103	101	103
Wavelet	13.85	16.79	0.54	0.29	0.97	0.98
Wavelet-ICA	14.23	17.65	0.47	0.21	0.98	0.99

In Figure 3, ECG signal with motion artifact, cleaned ECG signals using wavelet and wavelet-ICA methods and also noise estimated using these two methods are presented.

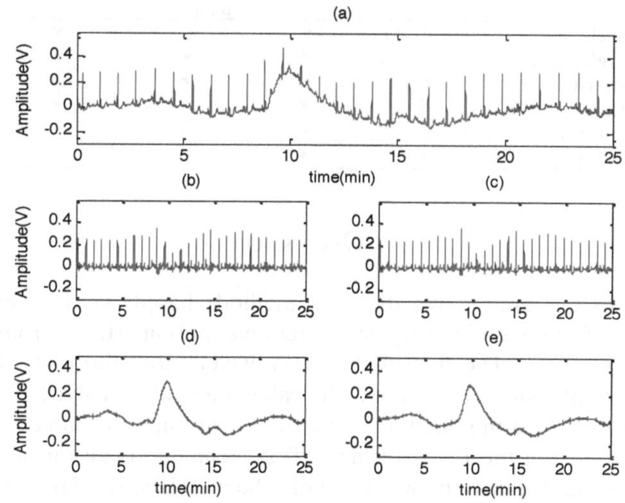

Fig. 3 (a) ECG signal with motion artifact, (b) cleaned ECG using wavelet, (c) cleaned ECG using wavelet-ICA, (d) noise estimated using wavelet and (e) noise estimated using wavelet-ICA

The PSD of ECG signal with motion artifact, clean signal and cleaned ECG signals using wavelet and wavelet-ICA are presented in Figure 4.

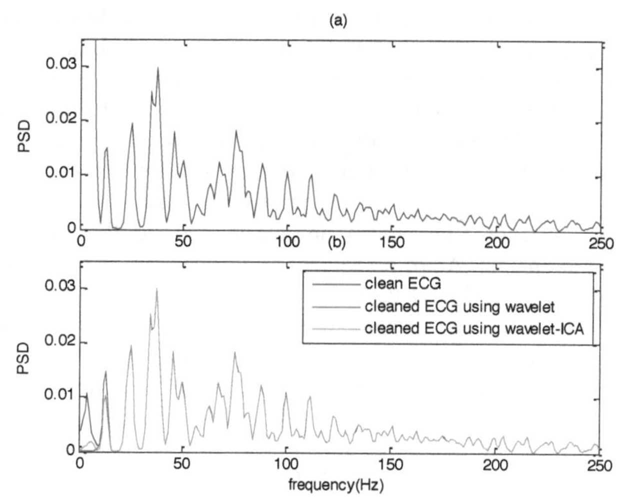

Fig. 4 PSD of (a) ECG signal with motion artifact, (b) clean ECG signal and cleaned ECG signal using wavelet and wavelet-ICA

Finally, the proposed methods were compared with other currently used methods such as adaptive filter, wiener filter [4], Kalman filter [4] and combined empirical mode decomposition (EMD) and ICA (EMD-ICA) [4] to remove motion artifact from ECG signals. The comparison results of these methods are shown in Table 2.

Table 2 Comparing different methods for noise removal from ECG signal

Methods	Automatic operation	On-line operation	Not requiring multiple input	Low computation burden
Wavelet	✓	✓	✓	✓
Wavelet-ICA	✓	✓	✓	✓
Adaptive filter	✓	✓	×	×
Wiener filter	✓	×	×	✓
Kalman filter	✓	✓	×	✓
EMD-ICA	×	×	✓	×

IV. Discussion

In this study, two different methods based on wavelet transform were investigated to remove motion artifact from ECG signal. The first method was wavelet transform based on nonlinear thresholding. The calculated values (SNR, RE and CC) for this method show this algorithm is capable of eliminating the ECG artifacts. This technique is an on-line method that can operate on single channel data. In Figure 3, PSD of cleaned ECG signal using wavelet transform shows this method is able to remove low-frequency components of noise from the distorted ECG signal. The next method that has been investigated was a combination method includes wavelet transform and ICA. This technique is an automatic artifact removal that does not need multi-channel signal. Also, it can operate on-line that is needed in many applications. As presented in Table 1, its performance is better than wavelet transform. However, the elapsed time for this method is more than the wavelet transform with nonlinear thresholding. Finally, the proposed methods were compared with other currently used methods such as adaptive filter, Wiener filter, Kalman filter and EMD-ICA (Table 2). According to this comparison, adaptive filter and Kalman filter are automatic methods that operate on-line [4]. However, they require additional sensor to remove noise. Wiener filter is an automatic method that cannot operate on-line. Also, it requires additional sensor for noise reference [4]. EMD-ICA is a single channel technique showed better performance compared to the wavelet based methods, but at the expense of a higher computational load. The difference in performance between the results of wavelet-ICA and EMD-ICA can be explained via differences in decomposing of the original signal. The computational load of EMD-ICA is high which makes a real-time implementation impossible. Hence, EMD-ICA cannot operate on-line [10]. Hence, to address the shortcomings of currently used methods as mentioned above, we proposed two different methods based on wavelet transform. These methods are able to operate automatic as we made wavelet-ICA an automatic method. Also, they operate on-line with a promising result. Furthermore, these methods do not require additional sensor for noise removal.

V. Conclusions

The experimental results show the capability of the proposed methods to remove motion artifact leaving behind the important information of the original signal. Also, it shows that the combination of wavelet transform and ICA method is able to produce better results compared to wavelet with nonlinear thresholding method.

Acknowledgment

The study was financed by the Knowledge Foundation's research profile Embedded Sensor System for Health (ESS-H).

References

[1] Strasser F, Muma M, Zoubir A M. (2012) Motion artifact removal in ECG signal using multi-resolution thresholding, EUSIPCO Proc., 20th European Conference on Signal Process., Bucharest, Romania, 2012, pp 899-903
[2] Hashim F R, Soraghan J, Petropoulakis L et al (2012) Wavelet based motion artifact removal for ECG signals, EMBS Proc., International Conference on Biomed. Eng. & Sci., Langkawi, Malaysia, 2012, pp 339-342
[3] Vishnubhotla R (2009) Pre-processing of ECG signals for ambulatory use. NIU Eng rev J 1:7-11
[4] Sweeney K T, Ward T E, McLoone S F (2012) Artifact removal in physiological signals-practices and possibilities. IEEE Trans Info Tech Biomed 16:488-500
[5] Abbaspour S, Fallah A (2014) A combination method for electrocardiogram rejection from surface electromyogram. Open Biomed Eng J 8:13-19
[6] Tealman J, Huffel S V, Spaepen A (2007) Wavelet-independent component analysis to remove electrocardiography contamination in surface electromyography, IEEE Proc., 29th Conference on Eng. Med. Biol. Soc., 2007, pp 682-685
[7] Daubechies I (1994) Ten lectures on wavelets, CBMS, SIAM
[8] Hyvärinen A, Oja E (2000) Independent component analysis: algorithms and applications. Neural Netw J 13:411-430
[9] He T, Clifford G, Tarassenko L (2006) Application of independent component analysis in removing artefacts from the electrocardiogram. Neural Comput Applic 15:105–116
[10] Taelman J, Mijovic B, Huffel S V et al (2011) ECG artifact removal from surface EMG signals by combining empirical mode decomposition and independent component analysis, BIOSIGNALS Proc., International Conference on Bio-inspired Syst. & Signal Process., Rome, Italy, 2011, pp. 421-424

An Affordable ECG and Respiration Monitoring System Based on Raspberry PI and ADAS1000: First Step towards Homecare Applications

F. Abtahi[1], B. Aslamy[1], I. Boujabir[1], F. Seoane[1,3], and K. Lindecrantz[1,2]

[1] School of Technology and Health, KTH Royal Institute of Technology, Stockholm, Sweden
[2] Department of Clinical Science, Intervention and Technology, Karolinska Institute, Stockholm, Sweden
[3] School of Engineering, University of Borås, Borås, Sweden

Abstract— **Homecare is a potential solution for problems associated with an aging population. This may involve several physiological measurements, and hence a flexible but affordable measurement device is needed. In this work, we have designed an ADAS1000-based four-lead electrocardiogram (ECG) and respiration monitoring system. It has been implemented using Raspberry PI as a platform for homecare applications. ADuM chips based on iCoupler technology have been used to achieve electrical isolation as required by IEC 60601 and IEC 60950 for patient safety. The result proved the potential of Raspberry PI for the design of a compact, affordable, and medically safe measurement device. Further work involves developing a more flexible software for collecting measurements from different devices (measuring, e.g., blood pressure, weight, impedance spectroscopy, blood glucose) through Bluetooth or user input and integrating them into a cloud-based homecare system.**

Keywords— **Raspberry PI, ECG, Respiration, ADAS1000,**

I. Introduction

An aging population—an increased percentage of elderly individuals in the overall population that has resulted from longer life expectancy and lower fertility rates—is challenging the current healthcare system by increasing costs, creating a lack of healthcare personnel, and contributing to more complex combinations of chronic diseases [1]. Homecare for elderly and chronic patients based on recent advances in information technology is a potential solution not only for decreasing the costs but also for increasing patients' quality of life. Potential homecare applications typically involve physiological and biological measurements such as blood pressure, blood glucose, weight, body composition, activity monitoring, and electrical cardiac activity through ECG; hence, the specific sensors and analogue front-ends required for different measurements of each target patient group may vary a lot. This makes the development of devices for pilot studies of homecare costly and, in particular, time consuming.

The aim of this work was to design a flexible, affordable platform for developing the measurement devices needed in different homecare pilot projects. ECG has been widely applied in diagnosing and monitoring heart disease and even in monitoring the autonomic nervous system through heart-rate variability (HRV). The ECG indeed is the most popular measured bio-signal, and it is also used in sports-training systems. Therefore, ECG has been chosen as a first example when developing and evaluating the platform. Advances in microelectronics and systems on chips create new opportunities for developing affordable compact devices. For instance, the ADAS1000 (ADAS) [2] introduced by Analog Devices includes all the components necessary to perform a five-lead ECG digital conversion with respiration and pacemaker-artefact detection in a single chip. In this work, a modular affordable vital-sign monitoring system is designed based on ADAS and Raspberry PI [3], a credit-card-size single-board computer.

II. System design

Figure 1 shows the proposed system architecture. The system includes a set of ECG electrodes connected to the patient and ADAS, which is connected to Raspberry PI (RPI) using a general-purpose input/output (GPIO), isolation for patient safety and to protect the device from electrostatic discharges. The schematic and single-layer printed circuit board (PCB) was designed using Cadence OrCAD 16.6. PCBs are manufactured by milling method using an in-house CNC machine.

A. Hardware

1) *ADAS1000 Chip* is a relatively new analogue front-end for ECG and thoracic electrical bioimpedance (EBI) measurements with integrated pacemaker-artefact and lead-off detection. It provides the lead, vector, or electrode signals at a sample rate of up to 128 kHz. The chip is designed to simplify the signal acquisition and ensure the robustness and electrical safety of the ECG devices in accordance with IEC 60601. These features—the significantly low power and the compact size—make ADAS specifically suitable for portable battery-operated devices, bedside patient monitoring, portable telemetry, and home monitoring systems. It comes in nine models with three or five ECG channels and options for thoracic impedance (respiration), a pacemaker-

H. Mindedal and M. Persson (eds.), *16th Nordic-Baltic Conference on Biomedical Engineering*,
IFMBE Proceedings 48, DOI: 10.1007/978-3-319-12967-9_2

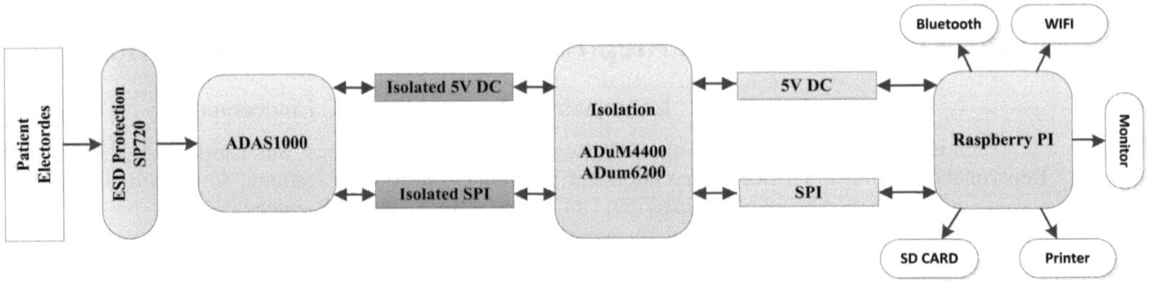

Figure 1.System Architecture: ADAS1000, Raspberry PI, defibrillation, and patient isolation. Signals can process in Raspberry PI, and results can be sent to a monitor or a printer or saved on an SD card; they can even sent by Bluetooth or Wi-Fi to another device or server.

artefact detection algorithm, and right-leg drive (RLD) input in low profile quad flat package (LQFP) or lead frame chip scale package (LFCSP) packages. Several ADAS chips can be combined as a master and slaves if more ECG channels are required. In this work, the ADAS1000BSTZ is used—that is, the five-channel version includes all previous features. The chip comes in a LQFP 64 pin package; the precision of our facilities for PCB assembly cannot handle such a tight grid; an LQFP adapter has therefore been used.

2) *Raspberry PI* is a credit-card-size single-board computer with an ARM processor. RPI is developed in two models, A and B, by Raspberry Foundation. Their aim is to facilitate the study of computer science and related topics through an affordable compact computer for students. However, the presence of a general purpose input/output (GPIO) makes it an ideal platform for developing standalone measurement systems. Model B, used in this project, provides Ethernet, two USB ports, HDMI, and audio and video output, in addition to a 700 MHz CPU, GPU, and 512 MB RAM. Storage can be provided by SD card or USB. RPI supports several Linux distributions, such as Raspbian (based on Debian and optimized for the Raspberry PI hardware) and even Android. Programming language is not limited by the operating system, and hence Python, Java, C/C++, and others can be used. In addition, it is possible to set up a web server such as Apache and a database such as MySQL. In general, RPI can be used as a platform in an affordable homecare setup. Bluetooth, WIFI, GSM/3G/LTE and even ZigBee can be used simply by employing USB dongles. And best of all, the RPI comes at a price of less than $80 (RPI, including enclosure, micro-USB power adapter, and high-speed SD card).

3) *SPI Communication through GPIO*, the serial peripheral interface (SPI) bus is developed by Motorola for full duplex communication between integrated circuits and on-board peripherals. SPI protocol uses four links to communication between one master and several slaves, a clock signal (SCLK) for synchronization, a slave select signal (SSn) and two data lines: master-out-slave-in (MOSI) and master-in-slave-out (MISO). Communication is controlled by the master, which selects the slave, activates the clock, and generates information on MOSI while it samples the MISO [4]. In this prototype, RPI acts as the master, communicating with a single ADAS (slave). However, it can easily expand by more ADAS chips to have more ECG leads.

4) *Defibrillator Protection* might be needed in any applications with possibility of defibrillation, such as in intensive care unit (ICU), protection against overvoltage is mandatory. In other fields of application it is still recommended, as it may protect the device from other types of electrostatic discharge (ESD). The ESD protection layer is designed according to the recommendations in the ADAS datasheet [2]. The protection circuit is based on SP724, which provides protection up to 8kV against ESD and other transient overvoltage events [5].

5) *Electrical Safety* is important in the design of any medical device. International electrotechnical commission (IEC) standards cover two types of insulation for patient protection, IEC 60601 and IEC 60950. The user (patient/operator) should be isolated from the high-voltage parts of the system, and leakage currents must be kept low. Otherwise, they may cause electric shock, cardiac arrhythmias, burns, or even damage to internal organs [6]. Isolation can be implemented at different levels, but it can also be ignored in applications with no connection, direct or indirect, to power-line voltage (e.g., in battery-driven holter devices). Since we cannot guarantee that this device will not connect to a power supply, printer, monitor, or other possible peripherals, we had to incorporate proper isolation into the design. Isolation is achieved by insulating both the SPI and DC power links between ADAS and RPI, as illustrated in Figure 1.

The SPI and DC lines are isolated by using the ADuM6200 and ADuM4400 from Analog Devices. They provide 5kV isolation, thus complying with IEC 60601 and 60950. The ADuM6200 provides isolated DC power, and ADuM4400 gives isolated digital communication, allowing

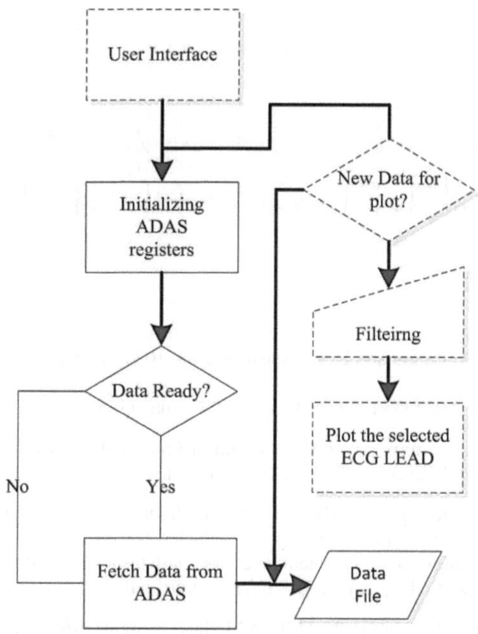

Figure 2.Simplified flowchart of software; boxes with dashes are implemented in Python, while solid ones are implemented in C.

a bit rate of 90 Mbps. Both chips are manufactured based on iCoupler technology [7], combining high-speed CMOS and monolithic air core transformer technology, which allows better cost, size, power, and reliability compared to optocouplers, which are common for isolation [8, 9].

B. Software

As mentioned before, RPI provides plenty of freedom of choice regarding operating system and programming language. In this case, Raspbian Linux and a combination of C and Python programming have been used. The implemented C software initializes the ADAS, retrieves the signals from ADAS, and records the signals in a file. Python is used to implement the display of the obtained signals in a parallel process. The reason for using two different programming languages is that C is very straightforward to use and has powerful libraries for GPIO communication, whereas plotting and visualizing the signals are more easily implemented in Python. A simplified flowchart of the software implementation is depicted in Figure 2.

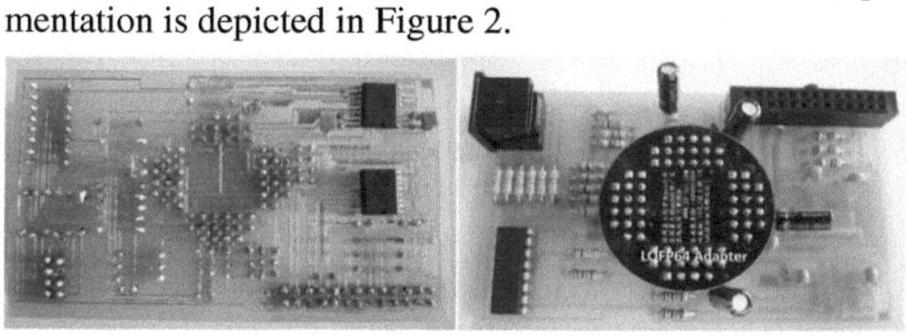

Figure 3.The top (right) and bottom (left) of assembled PCB. All components except ADuM4400 and ADuM6200 are assembled on Top.

III. RESULTS

The PCB board is shown in Figure 3. Table I summarizes the main components used in the design and their approximate costs. The total cost is about $170. The final circuit board is the same size as the RPI (94 × 57 × 1.5 mm), which is fairly compact as a standalone device. The evaluation of the measurement system has been performed using measurements from three healthy volunteers, and the designed device performed at the level of an ADAS evaluation board connected to a personal computer (PC). Figure 4 shows a 10 second epoch of a three-lead ECG and respiration for a healthy subject. Respiration is measured by EBI measurements at 56 kHz through Lead I. The thoracic impedance measurement for monitoring respiration can be done through Lead I, Lead II, or Lead III. It is clear that a respiration signal can be obtained from all three leads, but current injected through Lead I will pass both lungs; therefore, Lead I is the most suitable for monitoring respiration. As an alternative, it is possible to use an external path for EBI measurements and to place the electrodes across the thorax in order to achieve stronger respiratory influence on impedance measurements. ADAS uses two electrode configurations for EBI measurements (using the same pair of electrodes to inject the current and to measure the voltage); this method is sensitive to the skin-electrode impedance of electrodes, and it is not ideal for EBI measurements. On the other hand, tetra-polar EBI measurement is a robust method that reduces the influence of electrode impedance and skin-electrode contact impedance by using a pair of electrodes for current injection and another pair for voltage measurement. It is possible to use a simple analogue front-end to transform a two-electrode measurement system into a four-electrode measurement system, as introduced by Seoane et al. [10]. This can make ADAS useful also in impedance cardiography solutions.

The implemented software functions have demonstrated the basic features of storage, illustration, and signal processing. Further software development will provide more controls over ADAS's setting registers and processing options, such as heart-rate variability analysis. In addition,

TABLE I. COMPONENT COST OF FINAL PROTOTYPE

Component	*Cost ($)[a]*
Raspberry Pi MODEL B Starter Kit	80
ADAS1000-5	32
ADuM4400	8
ADuM6200	10
SP720, 4kV diode array	4
LQFP64 Adapter for ADAS1000	2
Resistors and capacitors, Connectors and cables	34
Total Cost	**170**

ADAS's built in features (e.g., lead-off detection, pacemaker-artefact detection) will be exploited, and users can choose the lead and frequency (46.5 kHz to 64 kHz) for respiratory (thoracic EBI) measurement. The system can also be integrated into cloud-based architecture for homecare monitoring. Further development can miniaturize the system so that it can play a primary role in personalized home health-monitoring solutions. Advances in textile electronics and printed electronics have demonstrated good performances for dry textile-based electrodes and printed electronics [11, 12] which can be used as part of ultimate solution.

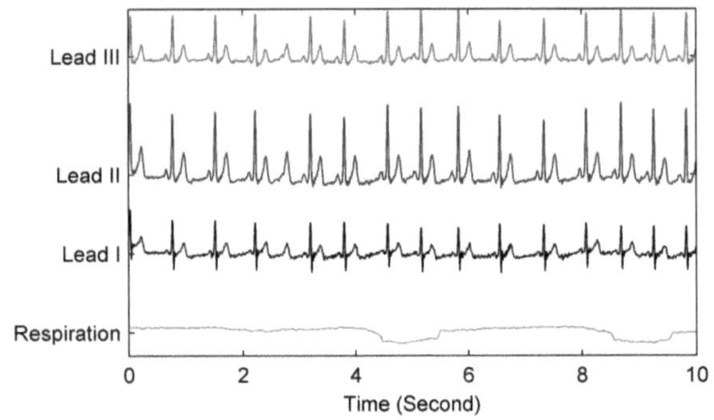

Figure 4.Respiration and ECG measurements, 10 seconds epoch.

IV. Discussion

The prototype uses a LQFP64 adapter for ADAS, which will be removed in next generations. In addition, further reduction in size is possible by using double layer instead of single layer PCB and surface-mount device (SMD) instead of through-hole components. The total cost of the prototype is less than $200, and higher volumes will yield lower prices. The cost of the final standalone platform is more affordable than other solutions based on PCs or even tablets. In addition, adding other physiological measurements like blood pressure, body weight, and body composition will make the solution more versatile and also economical for home monitoring. RPI has given us a flexible platform for further development. It is possible to add Bluetooth, Wi-Fi, GSM/3G, printers, camera by using a USB port or GPIO. It is worth mentioning that more USB ports can be added through a USB hub. It seems that powerful and affordable compact systems such as Raspberry PI, BeagleBone and Udoo have recently become direct competitors for microcontrollers in the development of flexible compact devices.

V. Conclusions

An ECG and respiration measurement device based on Raspberry PI and ADAS1000 as a flexible compact platform has been developed and evaluated. Further development of this platform can be part of a personalized health-monitoring and homecare system as a potential solution for the challenge of an aging population in the healthcare system.

References

1. Zweifel P, Felder S, Meiers M. Ageing of population and health care expenditure: a red herring? Health economics. 1999;8(6):485-96.

2. ADAS1000 Low Power, Low Power, Five Electrode Electrocardiogram (ECG) Analog Front End. Analog Devices; [cited 2014 10 Feb]; Available from: http://www.analog.com/static/imported-files/data_sheets/ADAS1000_1000-1_1000-2.pdf.

3. Pi R. An ARM GNU/Linux box for $25. Take a byte. 2012.

4. Leens F. An introduction to I 2 C and SPI protocols. Instrumentation & Measurement Magazine, IEEE. 2009;12(1):8-13.

5. SP724 Series 3pF 8kV Diode Array RoHS Datasheet. Littelfuse; [cited 2014 13 Feb]; Available from: http://www.littelfuse.com/~/media/electronics/datasheets/tvs_diode_arrays/littelfuse_tvs_diode_array_sp724_datasheet.pdf.pdf.

6. Krakauer D. Isolating the USB in Medical Systems. Medical Electronics Manufacturing [cited 2014 13 Feb]; Fall 2009:[Available from: http://www.devicelink.com/mem/archive/09/10/003.html.

7. Chen B. iCoupler® Products with isoPower™ Technology: Signal and Power Transfer Across Isolation Barrier Using Microtransformers. INTERFACE. 2006;1(C2):12mm.

8. ADuM4400 5 kV RMS Quad-Channel Digital Isolators Analog Devices; 2012 [cited 2014 10 Feb]; Available from: http://www.analog.com/static/imported-files/data_sheets/ADuM4400_4401_4402.pdf.

9. ADuM6200 Dual-Channel, 5 kV Isolators with Integrated DC-to-DC Converter. Analog Devices; 2012 [cited 2014 10 Feb]; Available from: http://www.analog.com/static/imported-files/data_sheets/ADuM6200_6201_6202.pdf.

10. Seoane F, Ferreira J, Sanchéz JJ, Bragós R. An analog front-end enables electrical impedance spectroscopy system on-chip for biomedical applications. Physiological measurement. 2008;29(6):S267.

11. Marquez JC, Seoane F, Valimaki E, Lindecrantz K, editors. Textile electrodes in electrical bioimpedance measurements–a comparison with conventional Ag/AgCl electrodes. Engineering in Medicine and Biology Society, 2009 EMBC 2009 Annual International Conference of the IEEE; 2009: IEEE.

12. Xie L, Yang G, Xu L, Seoane F, Chen Q, Zheng L, editors. Characterization of dry biopotential electrodes. Engineering in Medicine and Biology Society (EMBC), 2013 35th Annual International Conference of the IEEE; 2013: IEEE.

Author: Farhad Abtahi
Institute: Royal Institute of Technology (KTH
Street: Alfred Nobels Alle 10
City: Stockholm
Country: Sweden
Email: farhad.abtahi@sth.kth.se

Tailoring of HFPV Treatment by Respiratory Parameters Measurement

M. Ajčević[1], U. Lucangelo[2], and A. Accardo[1]

[1] Dept. of Engineering and Architecture, University of Trieste, Trieste, Italy
[2] Dept. of Perioperative Medicine, Intensive Care and Emergency,
Cattinara Hospital, University of Trieste, Trieste, Italy

Abstract— **High-frequency percussive ventilation (HFPV) is a non-conventional ventilatory modality which has proven highly effective in patients with severe gas exchange impairment. HFPV delivers a series of high-frequency sub-tidal volumes, by pulsatile flow, in combination with low-frequency breathing cycles. Currently, the mean airway pressure is the only parameter that allows an indirect comparison between conventional volume controlled ventilation and HFPV treatment set up. In the clinical practice, HFPV is not an intuitive ventilatory modality and the absence of valuable information during the treatment as delivered tidal volume, respiratory system resistance and compliance produces disaffection among the physicians. The aim of this study was to tailor HFPV treatment setup, avoiding under and overtreatment, using new bedside measurements of the main respiratory parameters. For this purpose pressure, flow and volume were measured in six patients undergoing HFPV. Respiratory system resistance and compliance were estimated using multiple linear regression method on Dorkin's high frequency model, while endotracheal tube pressure drop was estimated using Blasius's model.**

This study offers a new approach for HFPV treatment tailoring in accordance with protective ventilatory strategy and optimization of the alveolar recruitment. The possibility to measure respiratory parameters has opened a new horizon in clinical monitoring during this particular ventilator strategy, improving the beneficial effect of HFPV and minimizing iatrogenic lung damage.

Keywords— **HFPV, biomedical signal processing, personalized treatment.**

I. INTRODUCTION

Intubation and mechanical ventilation are life-saving therapies frequently used in cases of impaired gas exchange. High-frequency percussive ventilation (HFPV) is a ventilatory modality which associates the beneficial aspects of conventional mechanical ventilation with those of high-frequency ventilation [1]. This particular high frequency ventilation technique delivers a series of high-frequency sub-tidal volumes, by pulsatile flow, in combination with low-frequency breathing cycles. Over time HFPV was found highly effective in the treatment of several different respiratory system pathological conditions [2-4].

Currently, the HFPV ventilator VDR-4® (Volumetric Diffusive Respirator) displays the airway pressure waveform together with the corresponding peak and mean values. At the present time, the mean airway pressure, together with gas exchange analysis, are the only parameters that guide the physician in assessing the clinical effectiveness of HFPV and allow a partial comparison between conventional ventilation and HFPV treatment set up [5].

The evaluation of respiratory resistance and compliance parameters in patients undergoing conventional mechanical ventilation is currently used to detect lung dysfunctions and to evaluate the effect of treatment [6,7]. The end inflation occlusion method is used for the estimation of resistive and elastic characteristics of the respiratory system [6,7]. This method requires a constant flow, thus resulting inapplicable during HFPV, characterized by high frequency pulsatile flow. Recently, respiratory mechanics resistance and compliance parameters were estimated *in-vitro* during HFPV at different percussive frequencies [8] and at different working pressures [9] by using multiple linear regression approach based on Dorkin's high frequency model [10].

Furthermore, the airway pressure (Paw) measured by HFPV ventilator represents the sum of the endotracheal tube pressure drop (ΔP_{TT}) and the tracheal pressure (Ptr). From the clinical point of view, it is of paramount importance to take into account the real amount of pressure dissipated by endotracheal tube to avoid baro and volutrauma [11]. The recent *in-vitro* study identified the Blasius's model as the most adequate for the estimation of endotracheal tube pressure drop during this particular ventilation modality [12].

In the clinical practice, HFPV is not an intuitive ventilatory modality and the absence of valuable information during the treatment as delivered tidal volume, respiratory system resistance and compliance produces disaffection among the physicians. Avoiding the *volutrauma* is the cornerstone of the *protective ventilation strategy*, which requires a measurement of the inspiratory volume delivered to the patient. At the same time from a clinical point of view alveolar de-recruitment must be avoided, so *recruitment maneuvers* are frequently necessary to keep the lung open. Presently it is not possible to evaluate the delivered volume and the alveolar recruitment effect during HFPV treatment. Moreover, the systematic measurement of respiratory signals and parameters in patients undergoing HFPV has not yet been reported.

The aim of this study was to tailor HFPV treatment setup, avoiding under and overtreatment, using new bedside measurements of the main respiratory parameters.

H. Mindedal and M. Persson (eds.), *16th Nordic-Baltic Conference on Biomedical Engineering*,
IFMBE Proceedings 48, DOI: 10.1007/978-3-319-12967-9_3

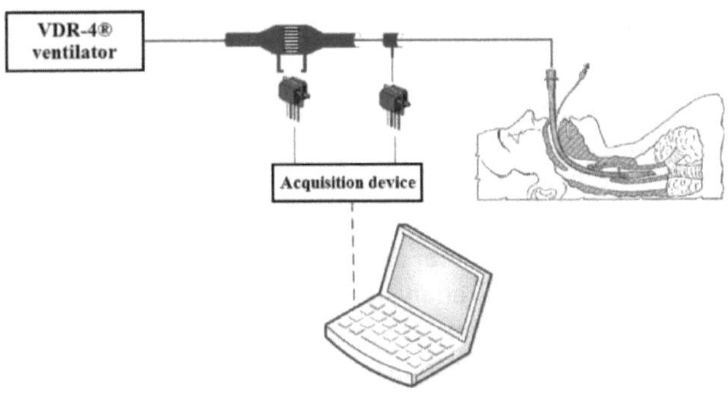

Fig. 1 Schematic diagram of respiratory circuit and acquisition system

II. MATERIALS AND METHODS

A. Acquisition system

The measurements of respiratory signals in patients undergoing HFPV were performed with a device suitably designed for bedside use. The pressure signal was measured by (ASCX01DN, Honeywell, USA) pressure transducer, while the flow signal was measured using Fleisch pneumotachograph (Type 2, Lausanne, Switzerland) linked to a differential pressure transducer (0.25 INCH-D-4V, All Sensors, USA). The signals were filtered with a low-pass second order Butterworth filter (cut-off frequency of 300 Hz) and sampled at 2 kHz by a 14 bit acquisition board (NI USB-6009, National Instruments, Austin, TX, USA) connected to a PC notebook. The volume and its acceleration were calculated by numerical integration and differentiation of flow, respectively.

The schematic diagram of respiratory circuit and acquisition system is shown in figure 1. The VDR-4® ventilator was connected to endotracheal tube of intubated patient through a Fleisch pneumotachograph and a sideport for pressure measurement. The acquisition device was supplied using isolated power supply, so there was no risk of electrical injury. In order to prevent water vapor condensation in the duct system, the apparatus was heated for five minutes before and during use, by small heating element built round the pneumotachograph.

B. Estimation of respiratory parameters

The estimation of the respiratory system resistance and compliance was performed by using multiple linear regression approach on Dorkin's model, which approximates the respiratory mechanics function at high frequencies [10]:

$$P_{aw}(t) = R \cdot \dot{V}(t) + \frac{1}{C} \cdot V(t) + I \cdot \ddot{V}(t) + P_0 \qquad (1)$$

where Paw(t) represents the pressure applied to the respiratory system, V(t) is the pulmonary volume, $\dot{V}(t)$ is the airflow, $\ddot{V}(t)$ is the volume acceleration and P_0 represents the pressure offset. The resistance R, compliance C and inertance I parameters represent respectively the viscous, elastic and inertial mechanical properties of the respiratory system.

The endotracheal pressure drop (ΔP_{TT}) was calculated using Blasius's flow dependent model:

$$\Delta P_{TT}(t) = K_B \cdot \dot{V}(t)^{1.75} + I_T \cdot \ddot{V}(t) \qquad (2)$$

where K_B e I_T are the resistance and inertance coefficients, respectively. The K_B and I_T used in this study were those previously obtained in [12].

C. Treatment protocol

HFPV was substituted for conventional volume controlled ventilation (VCV) in six patients who did not respond to conventional treatment after 12 hours. During conventional treatment tidal volume (V_T) delivery was set in accordance to protective ventilatory strategy, i.e. $V_T < 8$ mL/Kg predicted body weight (PBW) [13]. HFPV was set up similarly by using the same inspiratory-expiratory time ratio (I:E), mean airway pressure (Paw_{mean}) and positive end-expiratory pressure (PEEP). The percussive frequency of pulsatile flow was set to 500 cycles/min, and the pulse inspiratory and expiratory ratio (i:e) was fixed to 1. After a short stabilization period (10 minutes), the measurement of respiratory parameters pressure, flow and volume was performed and the R, C and ΔP_{TT} parameters were estimated.

In case the tidal volume was above the protective limit (i.e. V_T>8 mL/Kg_{PBW}), the flow was reduced on the VDR-4®. When the safety tidal volume was reached the measurement was repeated.

A *mini recruitment maneuver* was then applied in the form of peak airway pressure increment by the calculated ΔP_{TT} value. A new measurement of respiratory parameters was performed in order to assess the variations in respiratory system and verify if the tidal volume is within the safety limits.

During the following 12 hours the patient was ventilated maintaining these ventilator settings and every 4 hours blood samples were collected for gas analysis. The ratio between arterial pressure oxygen level and inspired oxygen fraction PaO_2/FiO_2 was calculated. At the end of 12 hours of HFPV treatment the patients returned to VCV.

This study was performed following a standardized HFPV clinical protocol without interfering with patients care and under the supervision of a physician not involved in the study. The study posed no added risk to the patient and did not interfere with usual patient care.

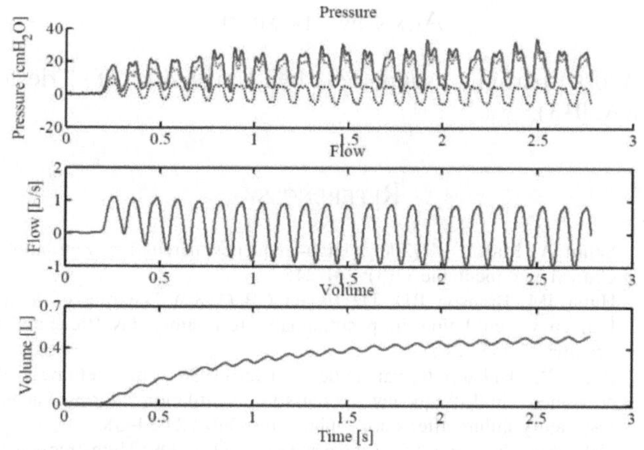

Fig. 2 From top to bottom: Pressure (airway pressure Paw – blue line, calculated tracheal pressure Ptr – red line, calculated ΔP_{TT} - black line), Flow and Volume during an inspiratory phase of a single respiratory cycle.

III. Results

Figure 2 shows the tracings of measured airway pressure, flow and volume, in a patient undergoing HFPV. In the top panel are also depicted calculated ΔP_{TT} (black dotted line) and calculated tracheal pressure $P_{tr} = P_{aw} - \Delta P_{TT}$ (red dotted line).

During the transition from conventional ventilation to HFPV, maintaining the same mean airway pressure, V_T varied significantly and in two of six patients exceeded the safety limit. In these two cases the tidal volume was immediately decreased to safety range and the measurements were repeated.

The measured tidal volume, and the estimated resistance and compliance values before (V_T, R, C) and after (V_T', R', C') the application of ΔP_{TT} increment are reported in table 1.

After the application of the *mini recruitment maneuver*, the respiratory system compliance and delivered tidal volume varied significantly, while the variation of the respiratory airway resistance was negligible. In all cases the $\Delta C\%$ was positive (from 3.7% to 34.1%) suggesting a potential alveolar recruitment. Compliance increment greater than 10%, which was retained clinically significant, was detected in four patients. In the other two patients the perturbation maneuver did not affected compliance. At the same time in all patients, the tidal volume values remained within the safety limits.

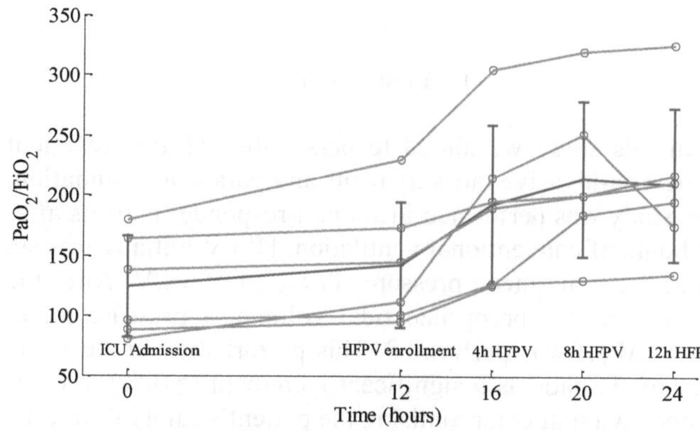

Fig. 3 PaO_2/FiO_2 values in the six patients at the ICU admission (initial observation), HFPV enrollment (after 12 hours), four, eight and twelve hours of HFPV treatment, respectively. Red line represents the mean and 1SD values among patients.

In figure 3, PaO_2/FiO_2 values, in six patients, before and during the HFPV treatment are reported. The gas exchange did not differ from initial observation to HFPV enrollment. After the first four hours of HFPV, PaO_2/FiO_2 increased significantly (mean slope = 12.6 h^{-1}). During the following eight hours of treatment in five of the six patients an increasing trend was detected. On the whole the mean slope of the 12 hours HFPV treatment was 5.6 h^{-1}, compared to 1.54 h^{-1} obtained during previous 12 hours of conventional treatment.

Table 1 Comparison between the respiratory parameters before and after the application of ΔP_{TT} increment.

Patient	R (cmH$_2$O/ L's)	C (mL/cmH$_2$O)	V_T/Kg_{PBW} (mL/Kg_{PBW})	ΔP_{TT} (cmH$_2$O)	R' (cmH$_2$O/ L's)	C' (mL/cmH$_2$O)	V'_T/Kg_{PBW} (mL/Kg_{PBW})	ΔC (%)
1	12	25	6.50	6	11.5	31	7.9	24
2	18.2	40.3	6.50	4	17.8	41.8	7.5	3.7
3	11	50.2	5.38	6	11.5	61.3	6.6	22.1
4	18	39	6.60	5	16.5	44	7.6	12.8
5	8.8	27.6	5.9	6	9.2	37	7.8	34.1
6	14	31	5.9	5	14	33	6.1	6.5

IV. Discussion

In this study we aimed to personalize HFPV treatment using non-invasive measurements and parameter estimation. The study was performed in six non-responder patients after 12 hours of conventional ventilation. HFPV initially was set up at the same mean pressure (Paw_{mean}) of VCV. After the assessment of appropriate tidal volume, a pressure increment (ΔP_{TT}) was performed. This perturbation produced in four of six patients a significant increment ($\geq$10%) of compliance without compromising the patient's safety (Table 1). This behavior is characteristic of alveolar recruitment and can be described as a shift from low inflation zone to linear zone in the pressure–volume curve, as described in [14]. The elastic lung properties can suddenly change when alveolar recruitment or de-recruitment occurs. From a clinical point of view alveolar de-recruitment must be avoided in order to keep the lung open.

On the contrary, the resistance did not change significantly after perturbation in all patients. In fact the airway resistance mainly depends on the presence of secretions and their mobilization is a known, but not immediate, feature of HFPV [15].

Blood gas analyses were recorded before and during the HFPV treatment and the PaO_2/FiO_2 ratio did not differ from initial observation to HFPV enrollment. During the first four hours of the HFPV treatment, after the mini recruitment maneuver, the mean slope of PaO_2/FiO_2 was 12.6 h^{-1}, compared to 10.2 h^{-1} obtained in the previous HFPV study based only on Paw_{mean} setup [5] without ΔP_{TT} compensation. This increment is clinically relevant particularly because obtained in accordance with the protective ventilatory strategy. However these preliminary results should be confirmed in future study.

V. Conclusions

This study offers a new approach for HFPV treatment tailoring in accordance with protective ventilatory strategy and optimization of the alveolar recruitment. The possibility to measure respiratory parameters has opened a new horizon in clinical monitoring during this particular ventilator strategy improving the beneficial effect of HFPV and minimizing iatrogenic lung damage.

Acknowledgment

Work partially supported by University of Trieste (FRA2013).

References

1. Salim A, Martin M (2005) High-frequency percussive ventilation. Critical care medicine 33(3): 241-245
2. Hurst JM, Branson RD, De Haven CB (1987) The role of high-frequency ventilation in posttraumatic respiratory insufficiency. J Trauma 27:236 –242
3. Reper P, Dankaert R, van Hille F et al. (1998) The usefulness of combined high-frequency percussive ventilation during acute respiratory failure after smoke inhalation. Burns 24:34–38
4. Velmahos GC, Chan LS, Tatevossian R et al. (1999) High-frequency percussive ventilation improves oxygenation in patients with ARDS. Chest 116:440–446
5. Lucangelo U, Zin WA, Fontanesi L et al. (2012) Early short-term application of high frequency percussive ventilation improves gas exchange in hypoxemic patients. Respiration 84:369-376
6. Bernasconi M, Ploysongsang Y, Gottfried SB et al. (1988) Respiratory compliance and resistance in mechanically ventilated patients with acute respiratory failure. Intensive Care Med 14:547-553
7. Rossi A, Gottfried SB, Higgs BD et al. (1985) Respiratory mechanics in mechanically ventilated patients with respiratory failure. J Appl Physiol 58:1849-1858
8. Riscica F, Lucangelo U, Ferluga M et al. (2011) In vitro measurements of respiratory mechanics during HFPV using a mechanical lung model. Physiological measurement 32(6): 637-48
9. Ajcevic M, Accardo A, Fornasa E et al. (2014) Estimation of Respiratory Mechanics Parameters During HFPV. IFMBE Proc. vol. 41, XIII Mediterranean Conference on Medical and Biological Engineering and Computing, Seville, Spain, 2013, pp 591–594
10. Dorkin HL, Lutchen KR, Jackson AC (1988) Human respiratory input impedance from 4 to 200 Hz: physiological and modeling considerations. J Appl Physiol 64:823-831
11. Rocco PRM, Zin WA (1995) Modelling the mechanical effects of tracheal tubes in normal subjects. Eur Respir J 8:121–126
12. Ajčević M, Lucangelo U, Ferluga M et al. (2014) In vitro estimation of pressure drop across tracheal tubes during high-frequency percussive ventilation. Physiological measurement 35(2):177- 188
13. De Campos T, et al. (2000) Ventilation with lower tidal volumes as compared with traditional tidal volumes for acute lung injury and the acute respiratory distress syndrome. The Acute Respiratory Distress Syndrome Network. N Engl J Med 342(18):1301-1308
14. Albaiceta GM, Blanch L, Lucangelo U. (2008) Static pressure-volume curves of the respiratory system: were they just a passing fad?. Curr Opin Crit Care 14(1):80-86
15. Hall JJ, Hunt JL, Arnoldo BD et al. (2007) Use of high-frequency percussive ventilation in inhalation injuries. J Burn Care Res 28(3):396–400

Author: Miloš Ajčević
Institute: University of Trieste, Dept. of Engineering and Architecture
Street: Via Valerio 10
City: Trieste
Country: Italy
Email: milos.ajcevic@phd.units.it

Tracking Performance of Several Combinations of Common Evaluation Metrics and Sub-pixel Methods

J. Albinsson[1], T. Jansson[2], and M. Cinthio[1]

[1] Department of Biomedical Engineering, Faculty of Engineering, Lund University, Lund, Sweden
[2] Clinical Sciences Lund, Biomedical Engineering, Lund University, and Medical Services, Skåne University Hospital, Lund, Sweden

Abstract— **Motion estimation in a series of consecutive images is used in a variety of areas, e.g. video compression and investigation of tissue characteristics and organ function in medical images. Several methods exist both for estimating motions on a pixel level, e.g. block-matching in which two blocks in consecutive images are compared by an evaluation metric, and on a sub-pixel level. In this paper, we have evaluated the tracking performance of all combinations between three evaluation metrics and eight sub-pixel estimation methods. The tracking performance of a sub-pixel method varies depending on the evaluation metric used. This indicates that a reported tracking performance for a sub-pixel estimation method can be significantly different when combined with another evaluation metric. Also there is a large variation in the time needed for the motion estimations depending primarily on the sub-pixel method used but also on the evaluation metric.**

Keywords— **Sub-pixel estimation, block-matching, motion estimation, ultrasound**

I. INTRODUCTION

Motion estimation in a series of consecutive images is used in a variety of areas, e.g. video compression and investigating tissue characteristics and organ function in medical images. For example in ultrasound where tissue motion measurements results in functional information on the tissue-of-interest, and have attracted attention for various applications such as evaluation of cardiac [1], vascular [2], and skeletal muscle [3] function. It has been shown that tissue motion measurements can not only provide new information about the tissue-of-interest [2], but can also provide prognostic information for patients having suspected cardiovascular disease, both with cardiac [4] and vascular applications [5].

Several methods for estimating motion in a series of digital images exist, e.g. optical flow and phase correlation. In this work, we have used a basic block-matching method implemented with a small search region. In order to estimate the motion in an area, a block representing the area is chosen and compared to every block in the search region with the block most similar with the original block assumed to depict the same object. To determine the similarity between two blocks, an evaluation metric, e.g. sum of absolute difference (SAD), sum of squared difference (SSD), or two dimensional normalized cross correlation (NCC), can be used.

However, most of the time, the length of the motion of an object between two successive images is not an exact number of pixels. Therefore, an estimation of the motion on a level of pixels is not enough as the relative error between estimated movement and true movement will, at least for short distances, be larger than acceptable. Several methods exist to determine the motion on a sub-pixel level. The methods can roughly be divided into three sub-groups: 1) analytically solved using the evaluation metric values as input, 2) interpolation of evaluation metric values, and 3) image interpolation. Both the first and second group uses the evaluation metric values in order to determine the sub-pixel position; values that differ depending on the evaluation metric used.

In motion estimations, two performance criterions are vital: estimation time and tracking error. Naturally the error should be kept as small as possible in order to give a correct estimation; but for real-time applications also the estimation time is critical. Often, the size of the estimation error is negatively correlated to the estimation time.

We have found no previous study that evaluates the impact on the performance of a sub-pixel method when used together with different evaluation metrics. The aim of this work was to evaluate the performance of eight sub-pixel position methods when combined with three different evaluation metrics *in silico*.

II. MATERIAL AND METHODS

A. Image sequences

The investigations in this paper have been made on simulated ultrasound images. In ultrasound images, horizontal direction is commonly denoted lateral while vertical direction is denoted axial; a convention that has been adopted in this paper.

The cineloops (50 images each, pixel density: 8.1x4.1 pixel/mm) were simulated using Field II [6]. The lateral velocities included 0.3, 0.6, 0.9, 1.4, 2.2, and 2.8 pixels/image whereas the axial velocity was 0 pixels/image. Motion estimation was performed on 100 blocks (15x7

H. Mindedal and M. Persson (eds.), *16th Nordic-Baltic Conference on Biomedical Engineering*,
IFMBE Proceedings 48, DOI: 10.1007/978-3-319-12967-9_4

pixels each) in each image using a full-search scheme (11x11 pixels) centered on the pixel position of the searched-for block. The searched-for blocks were evenly distributed above and below the focus depth.

B. Evaluation metrics

The performance for motion estimation was evaluated for three evaluation metrics:

SAD $$\alpha = \sum_{i=1}^{i=l}\sum_{j=1}^{j=k}\left|X_{i,j} - Y_{i+m,j+n}\right| \quad (1)$$

SSD $$\beta = \sum_{i=1}^{i=l}\sum_{j=1}^{j=k}\left(X_{i,j} - Y_{i+m,j+n}\right)^2 \quad (2)$$

NCC $$\gamma = \frac{\sum_{i=1}^{i=l}\sum_{j=1}^{j=k}(X_{i,j}-\bar{X})(Y_{i+m,j+n}-\bar{Y})}{\sqrt{\sum_{i=1}^{i=l}\sum_{j=1}^{j=k}(X_{i,j}-\bar{X})^2\sum_{i=1}^{i=l}\sum_{j=1}^{j=k}(Y_{i+m,j+n}-\bar{Y})^2}} \quad (3)$$

Here α, β, and γ denotes the calculated evaluation metric value; m and n denotes the displacement between the two blocks; l and k denotes the size of the blocks; $X_{i,j}$ and $Y_{i,j}$ denotes the pixel values at position (i, j) in the tracked block and the compared block, respectively, while $\bar{X}$ and $\bar{Y}$ denotes the average pixel values of the respective blocks.

C. Sub-pixel estimation methods

A total of eight sub-pixel position methods were evaluated. Three of the methods were analytically solved: 1. 1D parabolic interpolation (1D PI), 2. grid slope interpolation (GS) [7], and 3. 2D parabolic interpolation (2D PI). Three of the methods interpolated the calculated evaluation metric values to a factor 128:1: 4. 1D PI, 5. 2D PI, and 6. 2D cubic interpolation (2D Cubic). Two methods interpolated the image data a factor 128:1: 7. 2D PI and 8. 2D Cubic.

1. 1D PI fits a one dimensional second-degree polynomial (4), to three evaluation metric values with the center value corresponding to the block with the best similarity to the searched-for block.

$$y = ax^2 + bx + c \quad (4)$$

The polynomial was fitted separately laterally and axially in order to get a sub-pixel estimation in both directions. Fitting + (4) gave, when analytically solved, the sub-pixel estimation:

$$\Delta x = \frac{\alpha_1-\alpha_3}{2(\alpha_1+\alpha_3-2\alpha_2)} \quad (5)$$

Here α_1, α_2 (center), and α_3 were evaluation metric values and Δx was the sub-pixel part of the movement with a value less than ±0.5 pixels.

2. GS estimates the sub-pixel position separately laterally and axially by use of two evaluation metric values from the current image: the center value corresponding to the block with the best similarity to the searched-for block and the value closest to zero of the two values next to the center value (left or right for lateral or up or down for axial estimation). GS also uses an evaluation metric value calculated between the searched-for block and a block in the same image as the searched-for block at the position of the evaluation metric value used in the current image:

$$\Delta x = 0.5\left(1 - \frac{\alpha_2-\alpha_i}{\alpha_{2,0}-\alpha_{i,0}}\right) \quad (7)$$

Here α_2 (center) and α_i were evaluation metric values in the current image and $\alpha_{2,0}$ and $\alpha_{i,0}$ were evaluation metric values in the previous image. It should be noted that one of $\alpha_{2,0}$ and $\alpha_{i,0}$ were zero [7].

3. In 2D PI the second-degree polynomial was two dimensional.

$$z = a + bx + cy + dx^2 + exy + fy^2 \quad (6)$$

The polynomial was fitted to nine evaluation metric values with the value in the center corresponding to the block with the best similarity to the searched-for block. The polynomial was then solved analytically by finding the extreme point close to the center position.

4. The polynomial (4) was used for interpolating evaluation metric values in 127 evenly distributed points between each pixel position before finding the position with the best evaluation metric value. This was done separately axially and laterally.
5. The polynomial (6) was used for interpolating evaluation metric values in 127 evenly distributed points between each pixel position both axially and laterally before finding the position with the best evaluation metric value.
6. 2D Cubic used cubic spline interpolation [8] in order to interpolate nine evaluation metric values to 128:1 both axially and laterally. The optimal position was given directly by the best evaluation metric value.
7. The polynomial (6) were fitted to nine pixel values and were used to interpolate a square, ±0.5 pixels both laterally and axially, around the center pixel to 128x128 samples for a total of 15x7 original pixels before performing a full-search with the chosen evaluation metric.
8. 2D Cubic used cubic spline interpolation [8] in order to interpolate 15x7 pixel values to 128:1 both axially and laterally original pixels before performing a full-search with the chosen evaluation metric.

D. Evaluation of tracking performance

The tracking performance of each combination of evaluation metric and sub-pixel estimation method was evaluated by measuring the estimation time and by calculating the difference between the set movement and the estimated movement. The axial and lateral estimation errors were treated separately. The errors of the motion estimation (pixel/image) for each combination of evaluation metric, sub-pixel position method, and velocity are presented as boxplots in Figure 1. Values outside ±0.25 pixel are part of the statistics but not shown in the figure. The average estimation time in seconds needed for each image (100 blocks) is presented between the lateral and axial errors.

III. RESULTS AND DISCUSSION

Figure 1 shows average estimation time between the lateral and axial estimation errors. For every combination of evaluation metric and sub-pixel estimation method, the estimation errors have been combined in a boxplot. The results show an expected trade-off between estimation time and size of estimation errors. The smallest estimation error is in general obtained when the image was interpolated whereas the analytically solved methods are up to 150 times faster.

Reading the description of GS, we find nothing indicating that the method was tested or even intended to be used for determining sub-pixel positions axially. It could be hypothesized that the method are sensitive for noise when the distances between samples are short. This should explain the difference in the variance of estimation errors axially and laterally as the axially distance is a factor 2 shorter.

The large difference in tracking performance using 2D PI for image interpolation can be explained by discontinuities in the interpolated image, i.e. the interpolation will produce edges halfway between two original pixels where two polynomials meet. The edges will also cause a fluctuation of energy in the image. A great strength of NCC can be seen in Figure 1 as its tracking performance is significantly better for this sub-pixel method than the other evaluation metrics.

A tendency that can be observed in the results of several sub-pixel methods is the presence of bias in the tracking errors, i.e. the error is dependent on the size of the movement in the images.

However, several combinations give satisfying results both in terms of estimation error magnitude and estimation time. The figure also indicates, for the first time, possible combinations of methods for improved performance, e.g. using SAD and analytically solved 1D PI (axial)/GS(lateral).

Though the investigations in this paper have been made on ultrasound images, we are confident that similar results would be seen in clinical images obtained with other modalities such as MRI or CT.

IV. CONCLUSIONS

It is well known that the choice of evaluation metric can have a significant impact on both estimation time and the magnitude of the estimation error when using block-matching for motion estimation. Here, we show that a combination of evaluation metric and sub-pixel estimation method have an effect on the size of the estimation error. Thus, even if two evaluation metrics have the same tracking performance on a pixel level, the total tracking performance can differ depending on what sub-pixel estimation method is applied.

ACKNOWLEDGMENT

This study was supported by grants from the Swedish Research Council

REFERENCES

[1] W. N. McDicken, G. R. Sutherland, C. M. Moran, and L. N. Gordon, "Colour Doppler velocity imaging of the myocardium," Ultrasound Med. Biol., vol. 18, pp. 651-654, 1992.
[2] M. Cinthio, Å. R. Ahlgren, J. Bergkvist, T. Jansson, H. W. Persson, and K. Lindström, "Longitudinal movements and resulting shear strain of the arterial wall," Am. J. Physiol.-Heart. C., vol. 291, pp. H394-H402, 2006.
[3] S. Brorsson, A. Nilsdotter, M. Hilliges, C. Sollerman, and Y. Aurell, "Ultrasound evaluation in combination with finger extension force measurements of the forearm musculus extensor digitorum communis in healthy subjects," BMC Med. Imaging, vol. 8, p. 6, 2008.
[4] C. F. Madler, N. Payne, U. Wilkenshoff, A. Cohen, G. A. Derumeaux, L. A. Pierard, J. Engvall, L. A. Brodin, G. R. Sutherland, and A. G. Fraser, "Non-invasive diagnosis of coronary artery disease by quantitative stress echocardiography: optimal diagnostic models using off-line tissue Doppler in the MYDISE study," Eur. Heart. J., vol. 24, pp. 1584-1594, 2003.
[5] J. Blacher, A. P. Guerin, B. Pannier, S. J. Marchais, M. E. Safar, and G. M. London, "Impact of aortic stiffness on survival in endstage renal disease," Circulation, vol. 99, pp. 2434-2439, 1999.
[6] B. J. Geiman, L. N. Bohs, M. E. Anderson, S. M. Breit, and G. E. Trahey, "A novel interpolation strategy for estimating subsample speckle motion," Phys. Med. Biol., pp. 1541-1552, 2000.
[7] R. G. Keys, "Cubic Convolution Interpolation for Digital Image Processing," IEEE Trans. Acoust., Speech, Signal Process., vol. ASSP-29, pp. 1153-1160, 1981.
[8] J. A. Jensen, "Field: A program for Simulating Ultrasound Systems," 10th Nordic-Baltic Conference on Biomedical Imaging Published in Medical & Biological Engineering & Computing, vol. 34, pp. 351-353, 1996.

Corresponding author:

Author: John Albinsson
Institute: Biomedical Engineering
Street: Ole Römers väg 3
City: Lund
Country: Sweden
Email: john.albinsson@bme.lth.se

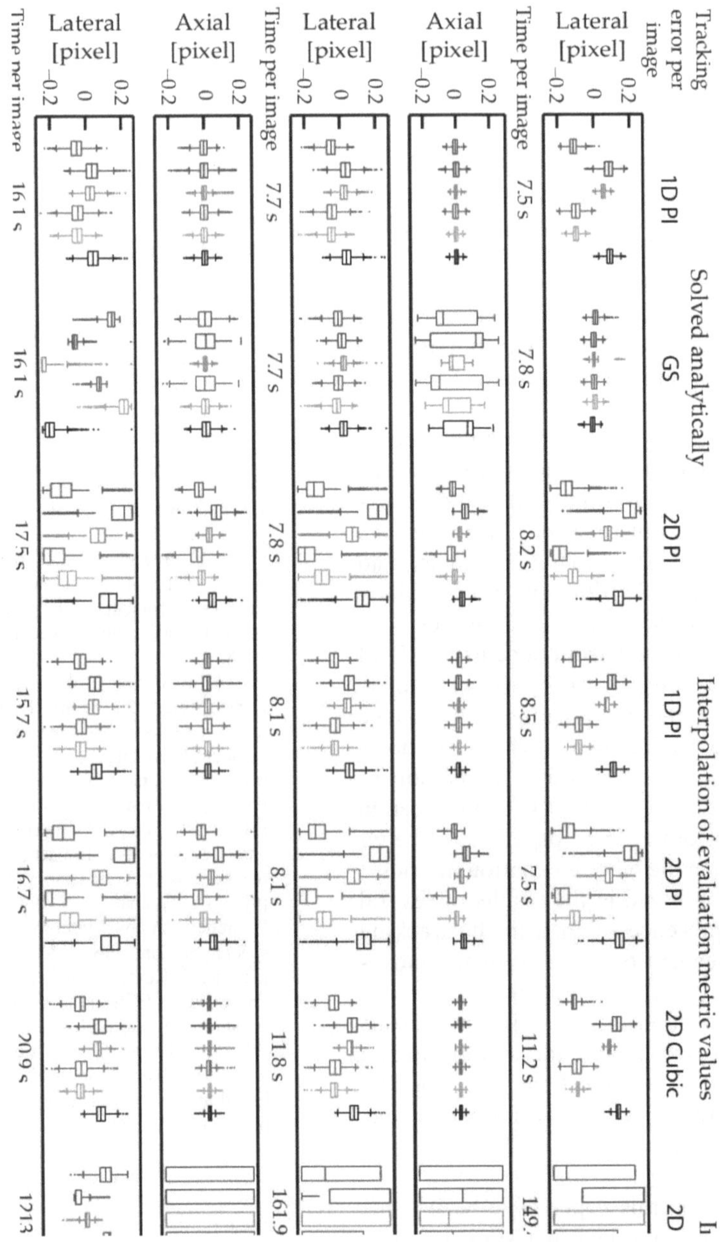

Fig. 1 Tracking error per image of the motion estimation performed on the in silico cineloops against the set movement per image of the cineloops. The boxes indicate the lower and upper quartiles and the median. The bar line indicate 99% of the values. Outliers are indicated as points. Each box is based on 4,900 estimations explaining the seemingly large numbers of outliers. No error larger than 0.25 pixels is shown but they were part of all the statistics.

Partial Volume Correction of Cerebral Perfusion Estimates Obtained by Arterial Spin Labeling

K. Ambarki[1,2], J. Petr[3], A. Wåhlin[1,4], R. Wirestam[5], L. Zarrinkoob[4], J. Malm[4], and A. Eklund[1,2]

[1] Department of Radiation Sciences, Umeå University, Umeå, Sweden
[2] Centre for Biomedical Engineering and Physics, Umeå University, Umeå, Sweden
[3] PET Center, Institute of Radiopharmaceutical Cancer Research, Helmholtz-Zentrum Dresden-Rossendorf, 01328 Dresden, Germany
[4] Center for Functional Brain Imaging, Department of Clinical Neuroscience, Umeå University, Umeå, Sweden
[5] Department of Medical Radiation Physics, Lund University, Lund, Sweden

Abstract— **Arterial Spin labeling (ASL) is a fully non-invasive MRI method capable to quantify cerebral perfusion. However, gray (GM) and white matter (WM) ASL perfusions are difficult to assess separately due to limited spatial resolution increasing the partial volume effects (PVE). In the present study, ASL PVE correction was implemented based on a regression algorithm in 22 healthy young men. PVE corrected perfusion of GM and WM were compared to previous studies. PVE-corrected GM perfusion was in agreement with literature values. In general, WM perfusion was higher despite the use of PVE correction.**

Keywords— **MRI, cerebral perfusion, gray and white matter, arterial spin labeling and healthy**

Introduction

Cerebral perfusion (CP, mL/min/100g) is of upmost importance to brain function. CP values can help clinicians in the diagnosis of different brain diseases such as vascular dementia, Alzheimer and stroke [1] [2]. For more than six decades [3], CP measurement has been an active technical challenge, which has lead to the development of a number of perfusion imaging techniques.

With the improvement of MRI perfusion techniques, it is now possible to assess, in a fully non-invasive fashion, CP using arterial spin labeling (ASL). ASL employs arterial water protons as an endogenous label by using spin inversion [4]. The ASL images have to be acquired during a relatively short time period limited by the arrival time of the labeled blood to the tissue and by T1 relaxation of the labeled blood. Also, the perfusion-weighted signal obtained as a difference between the control and labeled images has an amplitude in order of few percent of the control image. The measurement is thus typically repeated 20-40 times to reach good SNR. For the same reason the spatial resolution is kept around 3x3x5 mm3 to ensure whole-brain coverage is achieved within reasonable scan time

Today, using a 3D pseudo-continuous ASL scheme, gray matter (GM) and white matter (WM) perfusion can be assessed in less than 5 minutes, with full brain coverage.

However, fast imaging always results in images with limited or poor spatial resolution, which increases the partial volume effects (PVEs). PVE means that the ASL CP value within a certain voxel represents a combination of signal components from GM, WM and cerebrospinal fluid (CSF). This can easily be observed in the cortical GM with a thickness of about 2 mm and also by the subarachnoid CSF which leads to underestimation of the true value of GM perfusion [5] [6]. Recent studies have developed methods to correct the for PVEs in ASL data [7] [8].

In the present study, we aimed to implement PVE correction and used morphological transformation to assess the global GM and WM perfusion. This study was performed in healthy young men and in order to evaluate the method, our results were compared to CP values reported in the literature.

Method

Subjects

Twenty-two healthy young men were included in this study (26±2 years). None of the subjects had any sign of previous neurological disease, hypertension, cardiac disease, renal disease or peripheral vascular disease. The study was approved by the ethics review board of Umeå University (IRB).

Magnetic Resonance Imaging

All MRI data were collected at a 3 T unit (General Electric HD750 system) with a 32-channel head coil.

Structural data were obtained using a T1-weighted sagittal 3D fast spoiled gradient echo sequence to image the whole-brain with 176 slices, slice thickness 1 mm, field of view (FOV) 250×250 mm; acquisition matrix 256×256. These image data were used to create brain masks for GM and WM.

ASL data were collected using a 3D pseudo-continuous ASL (pCASL) method with an interleaved 3D stack of spiral fast spin echo readout sequence with background suppression [9]. The parameters were: 240 × 240 mm FOV, reconstruction matrix 128 × 128, ST 4 mm, 40 slices covering the whole-brain. CBF maps (mL/min/100g) were computed based on a single-compartment model using the following equation [10]:

H. Mindedal and M. Persson (eds.), *16th Nordic-Baltic Conference on Biomedical Engineering*,
IFMBE Proceedings 48, DOI: 10.1007/978-3-319-12967-9_5

$$CBF = \lambda\left(1-e^{-T_{sat}/T_{1GM}}\right)\frac{e^{w/T_{1GM}}}{2\alpha T_{1b}\left(1-e^{-\tau/T_{1b}}\right)}\frac{\Delta S}{S_0}$$

where τ is the labeling duration (1500 ms); α is the labeling efficiency (0.6) and w is the post-labeling delay (1525ms); λ is the tissue-to-blood partition coefficient (0.9); T_{1b} and T_{1GM} are T1 of blood (1600 ms) and gray matter (1200 ms), respectively, T_{sat} is the saturation time (2000 ms) of the reference image; S_0 and ΔS are the reference image and the ASL difference image, respectively. WM perfusion was computed using the same model and parameters.

BRAIN SEGMENTATION

MRI analysis was performed on de-identified images using the VBM8 toolbox (http://dbm.neuro.uni-jena.de/vbm.html) and the SPM8 software (http://www.fil.ion.ucl.ac.uk/spm/) with Matlab R2013b. Tissue compartments were classified into GM, WM and CSF (Fig. 1).

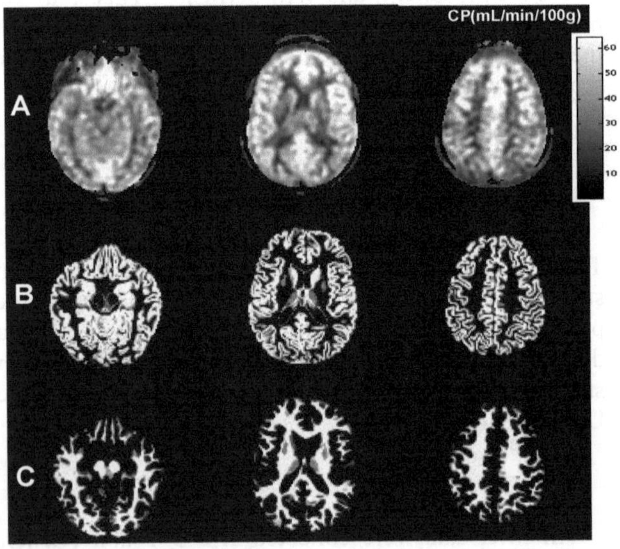

Figure 1 Cerebral perfusion by arterial spin labeling (A) and co-registered gray matter mask (B) and white matter mask (C) in three different z locations.

IMAGE POST-PROCESSING OF ASL DATA

Using the SPM8 software, the GM and WM masks obtained from VBM8 were co-registered to the ASL data and down-sampled to the same spatial resolution as ASL (Figure 1). These derived masks provided the fraction of GM, WM and CSF volumes in each ASL voxel.

The fractional volumes of GM, WM and CSF in each voxel, estimated by down-sampling the tissue probabilistic maps from VBM8, were used to perform partial volume correction [8] of the ASL data. The value of CBF in each ASL *i* pixel can be expressed by the following equation:

$$CBF(i)=F_{GM}(i)CBF_{GM}(i)+F_{WM}(i)CBF_{WM}(i)+F_{CSF}(i)CBF_{CSF}(i)$$

where F_{GM}, F_{WM} and F_{CSF} are the volume fractions of GM, WM and CSF, respectively, and CBF_{GM}, CBF_{WM} and CBF_{CSF} are the PVE-corrected perfusion estimates in GM, WM and CSF, respectively.

This PVC method is based on a regression algorithm previously described by Asslani et al.[7]. The perfusion maps of GM, WM and CSF were estimated by solving a system of nine linear equations.

In addition, to estimate the normal range of ASL perfusion in "pure" GM and WM, we created new masks using morphological erosion for WM and morphological closing for GM [11]. Only voxels containing more than 70% of GM or WM were included for estimation of the mean reference values of CBF_{GM} and CBF_{WM} and the corresponding ratio CBF_{GM}/CBF_{WM}.

We compared the following methods:
Method 1: No PVE correction, morphological operation or volume fraction threshold was used.

Method 2: PVE correction, morphological operation and volume fraction (>70%) were used, as described above

RESULTS AND DISCUSSION

Method	CP_{GM} (mL/min/100g)	CP_{WM} (mL/min/100g)	CP_{GM}/CP_{WM}
1	51±8	43±7	1.34±0.04
2	58±10	34±5	1.69±0.07

Table. Cerebral perfusion values (mean±SD) of gray matter (CP_{GM}) and white matter (CP_{WM}) and their ratio, using Method 1 and 2

The GM perfusion and the ratio CP_{GM}/CP_{WM} increased by 14% and 46 %, respectively, from Method 1 to Method 2 (Table), and the WM perfusion decreased accordingly by 22%. PVC maps of GM and WM are shown in Figure 2.

Our CP_{GM} results, based on Method 2, are in agreement with literature values, where the reported mean CP_{GM} values ranged from 58 to 67 mL/min/100g [12] [13] [14]. However, our findings were also 37% higher than in the study by Xu et al. (58 versus 48 mL/min/100g) [15] and, on the other hand, our value was 33% lower than values reported by Asllani et al (58 versus 88 mL/min/100g) [5]. Our CP_{GM} values agreed well with the PET study by Leenders et al. (58 versus 61 mL/min/100g) and by Henriksen et al. (58 versus 59 mL/min/100g) [16] [17] .

In general, our WM perfusion results were higher compared with literature data, which are ranging from 20 to 27mL/min/100g [17] [6]. This discrepancy could be explained by the different methods used to assess WM perfusion (e.g., globally or in specific regions using ROI analysis). Moreover, WM perfusion is generally problematic using ASL due to low SNR and another important issue is that the CBF mathematical model and the post-labeling delay used in the present study are optimized to assess the perfusion in GM and not WM.

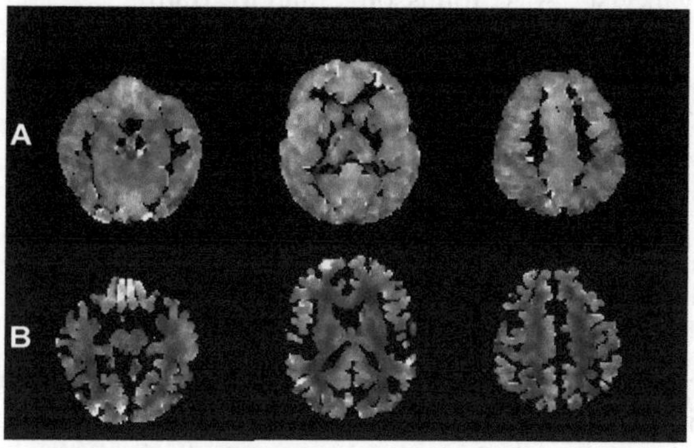

Figure 2 Partial volume corrected cerebral perfusion of gray (A) and white matter (B)

Conclusion

GM and WM perfusion levels were quantified in healthy young men. Partial volume correction seemed to improve the CP values of GM according to comparisons with literature values. However, CP values of WM were unreasonably high, despite the use of partial volume correction. Future work should evaluate the effects of post-labeling delay on ASL WM CP values in healthy individuals.

Acknowledgment

Swedish Research Council Grants 621-2011-5216 and 13514; European Union Objective 2 Norra Norrland (Project: 148273 CMTF); The County Council of Västerbotten and Swedish Heart and Lung Foundation Grant 20110383.

References

[1] R. Mielke, U. Pietrzyk, A. Jacobs, G. R. Fink, A. Ichimiya, J. Kessler, K. Herholz, and W. D. Heiss, "HMPAO SPET and FDG PET in Alzheimer's disease and vascular dementia: comparison of perfusion and metabolic pattern," *Eur J Nucl Med,* vol. 21, no. 10, pp. 1052-60, Oct, 1994.

[2] R. E. Latchaw, H. Yonas, G. J. Hunter, W. T. Yuh, T. Ueda, A. G. Sorensen, J. L. Sunshine, J. Biller, L. Wechsler, R. Higashida, G. Hademenos, and C. o. C. R. o. t. A. H. Association, "Guidelines and recommendations for perfusion imaging in cerebral ischemia: A scientific statement for healthcare professionals by the writing group on perfusion imaging, from the Council on Cardiovascular Radiology of the American Heart Association," *Stroke,* vol. 34, no. 4, pp. 1084-104, Apr, 2003.

[3] S. S. Kety, "Human cerebral blood flow and oxygen consumption as related to aging," *J Chronic Dis,* vol. 3, no. 5, pp. 478-86, May, 1956.

[4] M. Bernstein, K. King, and X. Zhou, *Handbook of MRI pulse sequences*: Elsevier Academic Press, 2004.

[5] I. Asllani, C. Habeck, A. Borogovac, T. R. Brown, A. M. Brickman, and Y. Stern, "Separating function from structure in perfusion imaging of the aging brain," *Hum Brain Mapp,* vol. 30, no. 9, pp. 2927-35, Sep, 2009.

[6] L. M. Parkes, W. Rashid, D. T. Chard, and P. S. Tofts, "Normal cerebral perfusion measurements using arterial spin labeling: reproducibility, stability, and age and gender effects," *Magn Reson Med,* vol. 51, no. 4, pp. 736-43, Apr, 2004.

[7] I. Asllani, A. Borogovac, and T. R. Brown, "Regression algorithm correcting for partial volume effects in arterial spin labeling MRI," *Magn Reson Med,* vol. 60, no. 6, pp. 1362-71, Dec, 2008.

[8] J. Petr, G. Schramm, F. Hofheinz, J. Langner, and J. van den Hoff, "Partial volume correction in arterial spin labeling using a Look-Locker sequence," *Magn Reson Med,* vol. 70, no. 6, pp. 1535-43, Dec, 2013.

[9] W. Dai, D. Garcia, C. de Bazelaire, and D. C. Alsop, "Continuous flow-driven inversion for arterial spin labeling using pulsed radio frequency and gradient fields," *Magn Reson Med,* vol. 60, no. 6, pp. 1488-97, Dec, 2008.

[10] D. C. Alsop, and J. A. Detre, "Reduced transit-time sensitivity in noninvasive magnetic resonance imaging of human cerebral blood flow," *J Cereb Blood Flow Metab,* vol. 16, no. 6, pp. 1236-49, Nov, 1996.

[11] H. J. Mutsaerts, E. Richard, D. F. Heijtel, M. J. van Osch, C. B. Majoie, and A. J. Nederveen, "Gray matter contamination in arterial spin labeling white matter perfusion measurements in patients with dementia," *Neuroimage Clin,* vol. 4, pp. 139-44, 2013.

[12] F. Q. Ye, K. F. Berman, T. Ellmore, G. Esposito, J. D. van Horn, Y. Yang, J. Duyn, A. M. Smith, J. A. Frank, D. R. Weinberger, and A. C. McLaughlin, "H(2)(15)O PET validation of steady-state arterial spin tagging cerebral blood flow measurements in humans," *Magn Reson Med,* vol. 44, no. 3, pp. 450-6, Sep, 2000.

[13] K. Zhang, H. Herzog, J. Mauler, C. Filss, T. W. Okell, E. R. Kops, L. Tellmann, T. Fischer, B. Brocke, W. Sturm, H. H. Coenen, and N. J. Shah, "Comparison of cerebral blood flow acquired by simultaneous [(15)O]water positron emission tomography and arterial spin labeling magnetic resonance imaging," *J Cereb Blood Flow Metab,* May, 2014.

[14] J. Wang, D. J. Licht, G. H. Jahng, C. S. Liu, J. T. Rubin, J. Haselgrove, R. A. Zimmerman, and J. A. Detre, "Pediatric perfusion imaging using pulsed arterial spin labeling," *J Magn Reson Imaging,* vol. 18, no. 4, pp. 404-13, Oct, 2003.

[15] G. Xu, H. A. Rowley, G. Wu, D. C. Alsop, A. Shankaranarayanan, M. Dowling, B. T. Christian, T. R. Oakes, and S. C. Johnson, "Reliability and precision of pseudo-continuous arterial spin labeling perfusion MRI on 3.0 T and comparison with 15O-water PET in elderly subjects at risk for Alzheimer's disease," *NMR Biomed,* vol. 23, no. 3, pp. 286-93, Apr, 2010.

[16] O. M. Henriksen, H. B. Larsson, A. E. Hansen, J. M. Grüner, I. Law, and E. Rostrup, "Estimation of intersubject variability of cerebral blood flow measurements using MRI and positron emission tomography," *J Magn Reson Imaging,* vol. 35, no. 6, pp. 1290-9, Jun, 2012.

[17] K. L. Leenders, D. Perani, A. A. Lammertsma, J. D. Heather, P. Buckingham, M. J. Healy, J. M. Gibbs, R. J. Wise, J. Hatazawa, and S. Herold, "Cerebral blood flow, blood volume and oxygen utilization. Normal values and effect of age," *Brain,* vol. 113 (Pt 1), pp. 27-47, Feb, 1990.

Author: Khalid Ambarki
Institute: Umeå University
Street:
City: Umeå
Country: Sweden
Email: Khalid.Ambarki@vll.

Automated Estimation of In-plane Nodule Shape in Chest Tomosynthesis Images

J. Arvidsson[1], A. Chodorowski[2,3], C. Söderman[4], A. Svalkvist[1], Å.A. Johnsson[5,6], and M. Båth[1,4]

[1] Department of Medical Physics and Biomedical Engineering,
Sahlgrenska University Hospital, SE-413 45 Gothenburg, Sweden
[2] Signals and Systems, Chalmers University of Technology, Gothenburg, Sweden
[3] MedTech West, Sahlgrenska University Hospital, Gothenburg, Sweden
[4] Department of Radiation Physics at the Institute of Clinical Sciences,
Sahlgrenska University Hospital, SE-413 45, Gothenburg, Sweden
[5] Department of Radiology at the Institute of Clinical Sciences,
Sahlgrenska University Hospital, SE-413 45 Gothenburg, Sweden
[6] Department of Radiology, Sahlgrenska University Hospital, SE-413 45 Gothenburg, Sweden

Abstract— **The purpose of this study was to develop an automated segmentation method for lung nodules in chest tomosynthesis images. A number of simulated nodules of different sizes and shapes were created and inserted in two different locations into clinical chest tomosynthesis projections. The tomosynthesis volumes were then reconstructed using standard cone beam filtered back projection, with 1 mm slice interval. For the in-plane segmentation, the central plane of each nodule was selected. The segmentation method was formulated as an optimization problem where the nodule boundary corresponds to the minimum of the cost function, which is found by dynamic programming. The cost function was composed of terms related to pixel intensities, edge strength, edge direction and a smoothness constraint. The segmentation results were evaluated using an overlap measure (Dice index) of nodule regions and a distance measure (Hausdorff distance) between true and segmented nodule. On clinical images, the nodule segmentation method achieved a mean Dice index of 0.96 ± 0.01, and a mean Hausdorff distance of 0.5 ± 0.2 mm for isolated nodules and for nodules close to other lung structures a mean Dice index of 0.95 ± 0.02 and a mean Hausdorff distance of 0.5 ± 0.2 mm. The method achieved an acceptable accuracy and may be useful for area estimation of lung nodules.**

Keywords— **segmentation, nodule, chest tomosynthesis, dynamic programming**

I. Introduction

Tomosynthesis (TS) is a 3D imaging technique that uses multiple radiographic images taken from an X-ray source placed at several positions [1, 2]. It is similar to conventional Computed Tomography (CT) technique, except that a limited angular range of x-ray projections is used. The resulting TS volume represents a 3D distribution of x-ray attenuation. The limited range of projections makes TS attractive, as the radiation dose in is estimated to be only 2 % of the radiation dose in conventional chest CT [3].

In chest TS the reconstructed volumes are used for e.g. detection of pulmonary nodules [4]. The correct estimation of nodule size and shape is thus important in clinical practice. However, 3D segmentation of TS volumes is difficult due to the low depth resolution from the limited angular range of projections and limitations in the reconstruction techniques [5].

In this paper we focus on the reconstructed volumes containing pulmonary nodules, in particular on the central plane of the nodules. This plane is often the most accurately reconstructed plane and is therefore used by clinicians to infer the nodules properties. We propose a semi-automated segmentation method for detection of the nodule shape in the central plane. The method is based on finding the nodule's pixels belonging to the nodule edge using edge detection techniques and applying dynamic programming to find the location of the nodule boundary. The segmentation method is evaluated against a ground truth binary mask obtained from the simulated nodule.

II. Material and Methods

A. Nodule Simulation

In the experiments we used synthetic lung nodules created according to the simulation technique described by Svalkvist et al. [6, 7]. In this technique the nodules are created in two steps: first a large-scale spherical nodule is created with a given radius, and then a random number of additional spheres with shifted centers are created and added to the original main nodule. The simulated nodules were created using a resolution (voxel size) of $0.1\times0.1\times0.1$ mm^3 and resulted in a binary nodule mask with values 1 inside the nodule, and values 0 outside the nodule. An example of a simulated nodule is shown in Figure 1.

Fig. 1 Example of a simulated nodule, Radius of 5.9 mm, shape consisting of 24 spheres

H. Mindedal and M. Persson (eds.), *16th Nordic-Baltic Conference on Biomedical Engineering*,
IFMBE Proceedings 48, DOI: 10.1007/978-3-319-12967-9_6

B. Tomosynthesis System

The clinical images used in this study were taken using the TS system GE Definition 8000 X-ray system with VolumeRAD (GE Healthcare, Chalfont St. Giles, UK). The system collects 60 low-dose projection images within approximately 10 seconds. The detector position is fixed, whereas the X-ray tube performs a vertical movement relative to the standard orthogonal posteroanterior (PA) projection. The projection images are collected in the angular interval of -15 deg. to +15 deg. The focus-detector distance is 180 cm in the PA projection. The detector size is 2022 × 2022 pixels, with a pixel size of 0.2×0.2 mm^2. The focal spot size is 1.25×1.25 mm^2. Figure 2 shows the system geometry.

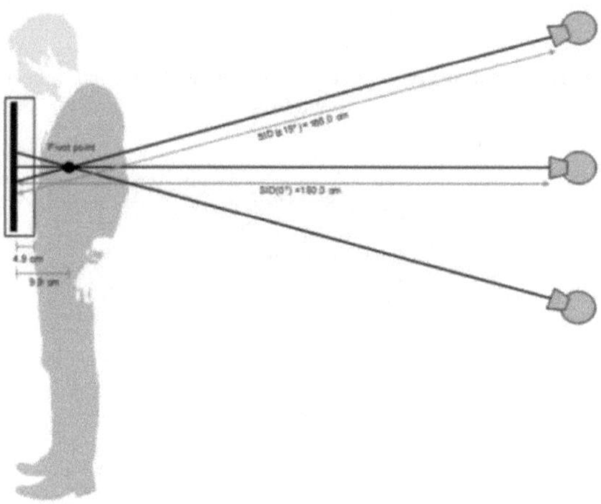

Fig. 2 Tomosynthesis system geometry

C. Nodule Insertion

The simulated nodules were inserted into the raw-data projection images, before reconstruction of the resulting section images. The nodules were positioned at the desired location in the 3D space and projected into the raw-data projection images of the TS acquisition, according to Svalkvist et al. [6, 7]. In this study a simplified insertion process was used in which the focal spot size, signal spread in the detector, scattered radiation and patient motion were not included. Two locations with different anatomy were used: one were the nodule was isolated from anatomical structures and one placed close to an anatomical structure (Fig. 3, 4). In the experiments, the nodule attenuation coefficient 0.02 $mm^{-1}$was used.

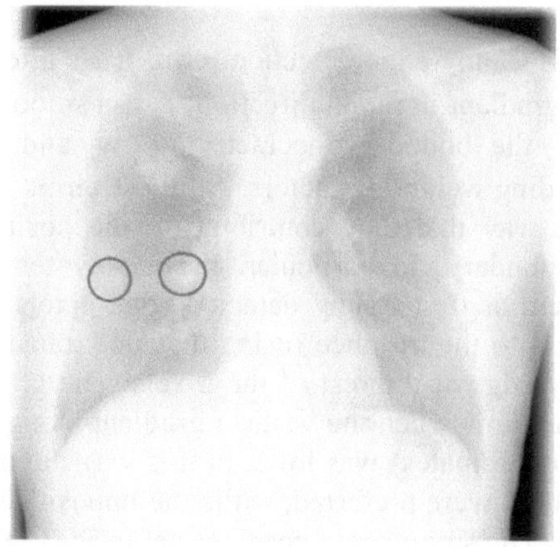

Fig. 3 Projection image with simulated nodules in isolated (left) and non-isolated (right) anatomical positions

D. 3-D Reconstruction

The TS reconstruction was performed using standard Cone Beam - Filtered Back Projection (CB - FBP), with an additional 3-D View Weighting correction technique described by Li et al. [8]. This technique corrects for intensity artifacts associated with CB – FBP by renormalizing voxel values based on the number of times that each voxel is modified during the back projection step in CB - FBP. The back projection filter was designed according to Lauritsch and Haerer [9].

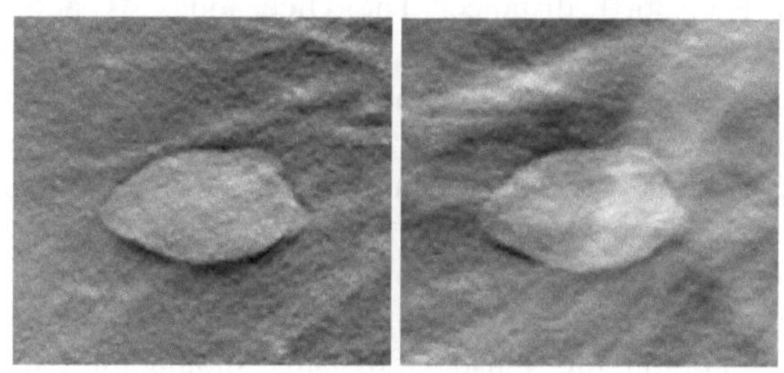

Fig. 4 Example of reconstructed nodule at the central slice, left: isolated nodule, right: nodule close to anatomical structure

E. Segmentation

The dynamic programming (DP) approach is frequently used for boundary detection in medical applications, in particular in segmentation of CT images [10, 11, 12]. The cost function used in DP should consist of terms that have impact on the position of the object boundary. In our case the local cost function $C(p,q)$ from pixel q to a neighboring pixel p is defined in polar coordinates as

$$C(p,q) = w_c \cdot f_c(p) + w_g \cdot f_g(p) + w_s \cdot f_s(p,q), \quad \text{(Eq. 1)}$$

where f_c is Canny edge detection term, f_g is a term related to image gradient in radial direction, f_s is a smoothness cost related to the boundary geometry, w_c, w_g and w_s are the corresponding weighting factors. All these terms are defined in such a way that they contribute to the position of the desired boundary. In particular, the Canny term function was defined as 0 at Canny detected edge-pixels and a real value equal to the distance (using distance transform) from the Canny edge at the rest of the pixels. The gradient term was defined low when the vertical gradient along the image (in polar coordinates) was low. In this way the edges from bright to dark were preferred, while the opposite edges were neglected. The smoothness cost was set to $|\Delta r|$, where Δr is a radius change for neighbouring boundary pixels. In this way the smooth and circular boundary is preferred, as large radial changes will be penalized. In addition, a constraint that nodule radius cannot be shorter than some predefined value (in our case 3 mm) was introduced. The weights in Eq. 1 denote the corresponding weighting factors, which can be set empirically or learned from the data. The center of the polar coordinate system was interactively defined by the user, by pointing on some central pixel inside the nodule.

F. Performance Evaluation

The segmentation accuracy was evaluated using Dice index and Hausdorff distance. The Dice index is defined as, $DI = 2 \cdot |S_a \cap S_e|/(|S_a| + |S_e|)$, where S_a and S_e are sets of pixels assigned to the object by the algorithm and the expert, respectively, and $|\ \ |$ is cardinality of the set. The Dice index varies between 0 and 1, where 1 denotes a perfect segmentation. The Hausdorff distance measures the extent to which each point on the segmented image lies near some point of the ground truth image [13]. In a perfect segmentation, the Hausdorff distance should approach 0.

III. RESULTS

Figures 5,6,7 demonstrate the results of image processing and image segmentation for one of the simulated nodules. In Figure 9 we present a performance evaluation of the developed method on a set of 18 simulated nodules with radius ranging from 2.9 to 9.9 mm. The Canny local cost term was implemented using the edge detection function from MATLAB's Image Processing Toolbox, using default parameter settings [15]. The Sobel filter was used as a gradient operator, minimum allowed radius was 3 mm and weighting factors were set to 0.5, 0.2, 0.2 for Canny cost, gradient cost and smoothness cost, respectively.

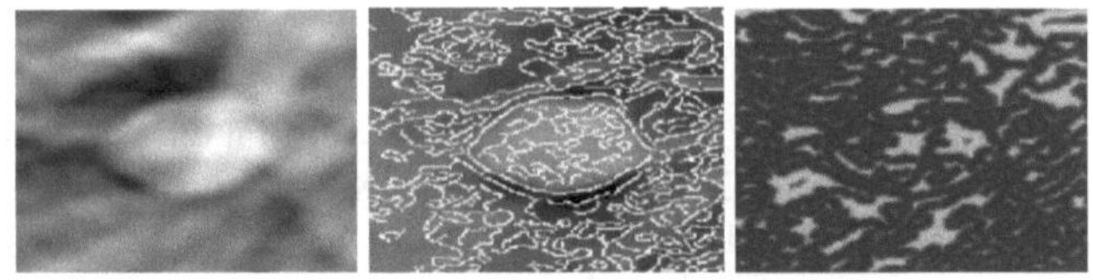

Fig. 5 Left: input image, middle: canny edges superimposed on the input image, right: canny edge cost image with distance transform

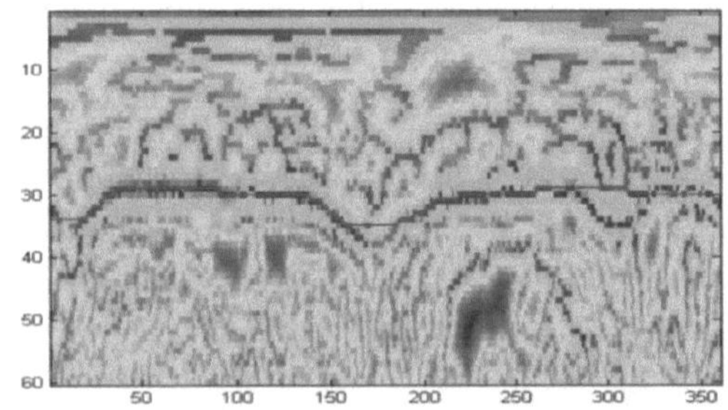

Fig. 6 Total cost image in polar coordinates with optimal path (in red) found by DP

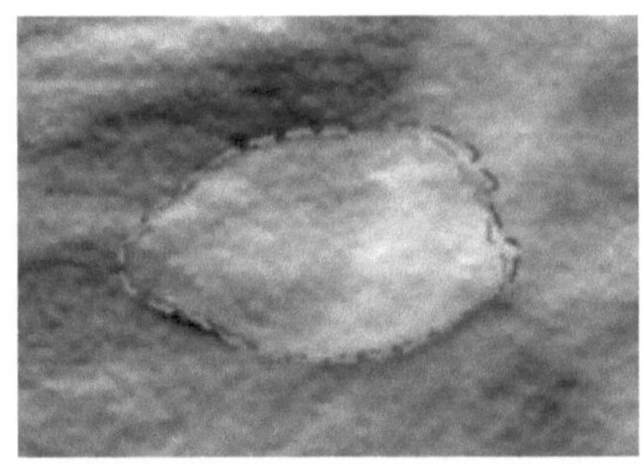

Fig. 7 Segmented boundary of nodule #11 (red), ground truth (green)

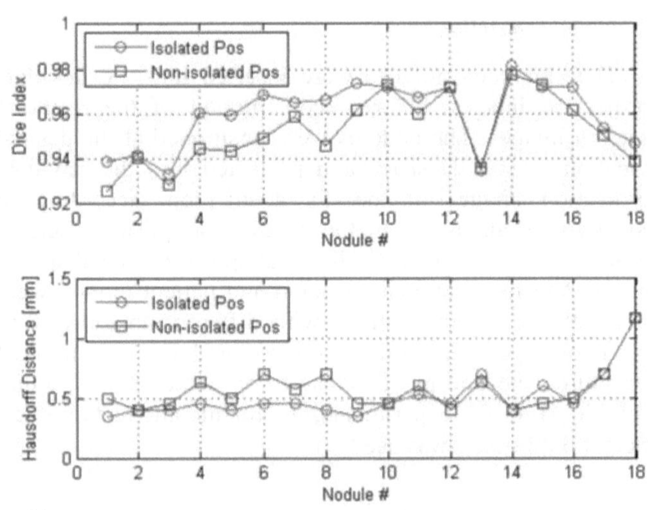

Fig. 8 Upper: Dice index, lower: Hausdorff distance evaluated on 18 simulated nodules with radius ranging from 2.9 to 9.9 mm, and two locations: isolated and non-isolated from anatomic lung structures. The nodules are in ascending order of size.

IV. DISCUSSION

In the present paper, a method of automated estimation of in-plane nodule shape in chest tomosynthesis images has been described. The results are promising and indicate that such segmentation could be feasible in the clinical environment. The method should be further evaluated on a larger dataset and by taking into account effects on nodule insertion such as system modulation function, scattering and patient motion. By disregarding these effects simulated data in which the nodule border appears slightly more distinct than for an equivalent real nodule are obtained.

It can be observed (Fig. 4) that filtered back-projection based reconstruction is associated with artifacts above and below homogenous structures. This artifact causes a false edge outside of the true nodule border and thus influences the results of the gradient based segmentation method.

In this work an in-house implementation of CB-FBP was used to reconstruct the TS data. Any improvements made to the reconstruction algorithm may also improve the segmentation performance.

TS data is typically associated with a long slice sensitivity profile i.e. low depth resolution. This gives rise to the question of how to evaluate the accuracy of the segmentation. Since this work is based on simulated nodules, ground truth is given from the corresponding slice of the simulated nodule. However, the low depth resolution implies that nodule structures from neighboring slices will be overlaid onto the central slice. This can potentially cause the nodule to appear larger and also influence the segmentation algorithm to over segment compared to ground truth. In this work the aim was to segment the central slice corresponding to that of the simulated nodule. In a clinical situation it may be of interest to instead mimic the segmentation performed by a clinical expert. In that case evaluation of the segmentation could instead be made against gold standard delineations made by clinical experts.

V. CONCLUSIONS

Accurate nodule shape estimation is important for nodule characterization in digital chest tomosynthesis. Our automated segmentation method achieved acceptable accuracy and may be useful for in-plane area estimation of lung nodules.

ACKNOWLEDGMENT

This work was supported by grants from the Swedish Research Council [2011/488, 2013/3477], the Swedish Radiation Safety Authority [2013/2982] and the Swedish Federal Government under the LUA/ALF agreement [ALFGBG-136281].

REFERENCES

1. Dobbins JT III (2009) Tomosynthesis imaging: at a translational crossroads. Med Phys 36:1956-1967
2. Dobbins JT III, McAdams H P (2009) Chest tomosynthesis: technical principles and clinical update. Eur J Radiol 72(2):244-51
3. Båth M, Svalkvist A, von Wrangel A et al. (2010) Effective dose to patients from chest examinations with tomosynthesis. Radiat Prot Dosimetry 139:153-158
4. Vikgren J, Zachrisson S, Svalkvist A et al. (2008) Comparison of chest tomosynthesis and chest radiography for detection of pulmonary nodules: human observer study of clinical cases, Radiology 249: 1034-1041
5. Warp R, Godfrey D, Dobbins J (2000) Applications of matrix inversion tomosynthesis, Medical Imaging 2000: Physics of Medical Imaging, Proc. SPIE 3977, pp. 1117-1125
6. Svalkvist A, Håkansson M, Ullman G et al. (2010) Simulation of lung nodules in chest tomosynthesis. Rad Prot Dosimetry 193:130-139
7. Svalkvist A, Johnsson Å A, Vikgren J et al. (2012) Evaluation of an improved method of simulating lung nodules in chest tomosynthesis. Acta Radiol 53:874-884
8. Li B, Avinash G, Claus B et al. (2006) 3-D View weighted cone-beam filtered backprojection reconstruction for digital tomosynthesis, RSNA Scientific Assembly and Annual Meeting Program, Chicago, IL., pp. 230
9. Lauritsch G, Haerer WH (1998) A theoretical Framework for filtered backprojection in tomosynthesis, Medical Imaging 1998: Image Processing, Proc. SPIE 3338, pp. 1127-1137
10. Wang J, Dobbins JT III, Li Q (2012) Automated lung segmentation in digital chest tomosynthesis, Med Phys. 39: 732–741.
11. Hu S, Hoffman E A, Reinhardt J M. (2001) Automatic lung segmentation for accurate quantitation of volumetric x-ray CT images. IEEE Trans on Medical Imaging, Vol. 20, No. 6, pp. 490-498
12. Sankar P V, Sklansky J (1982) A Gestalt-Guided Heuristic Boundary Follower for X-Ray Images of Lung Nodules, IEEE Trans on Pattern Analysis and Machine Intelligence, Vol.40, No.3, pp.326-331
13. Huttenlocher D P, Klanderman G, Rucklidge W J (1993) Comparing Images Using the Hausdorff Distance, IEEE Trans on Pattern Analysis and Machine Intelligence, Vol.15, No.9, pp.850-863
14. Canny J (1986) "A Computational Approach to Edge Detection", IEEE Transactions on Pattern Analysis and Machine Intelligence, Vol. PAMI-8, No. 6, pp. 679-698
15. MATLAB Image Processing Toolbox User's Guide, The Mathworks Inc., Natick, Massachusetts, www.mathworks.com , 2014

Information System for Postmarket Surveillance of Total Joint Prostheses

E. Berntsen[1] and Ankica Babic[1,2]

[1] Department of Information Science and Media Studies, University of Bergen, Bergen, Norway
[2] Department of Biomedical Engineering, Linköping University, Linköping, Sweden

Abstract—**Storage, integration and presentation of clinical data is an important aspect of any modern medical research. The Biomaterials research group at the Haukeland University Hospital uses both their own locally generated clinical data and external registry data to examine explanted joint implants. As a solution to this challenge, a system prototype was developed that would enable further integration of these information systems into a multi-user environment.**

The system allows importing registry data and matching it with local data, viewing and editing of this information and exporting the integrated data for further statistical analysis. An evaluation consisting of both user testing and heuristic evaluation was carried out and generated constructive feedback.

The prototype demonstrates the feasibility of combining these data sources in a single database and the future possibility of exposing parts of this information to external users through a web application.

Future integration of external sources could improve the information management of biobank data for postmarket surveillance of medical devices.

Keywords— **Data integration, postmarket surveillance, information sharing**

I. INTRODUCTION

The work that the system supports involves analyzing explanted knee implants that have failed and have been removed from the patients. The data generated form these analyses needs to be integrated with data about the patient and the implant itself. Today the researchers get this data from a national registry by requesting a SPSS file from a statistician at the registry. This is then imported to an adhoc Microsoft Excel spreadsheet on a network drive.

The current process is cumbersome and takes up a lot of the researcher's valuable time. It also has issues with concurrent access. Statistics from the registry show that the number of patients getting both hip and knee implants have risen considerably in the recent years [1]. Due to the general population of Norway living longer on average, and the relative large size of the post-war generation [2], the amount of patients needing joint implants is projected to rise dramatically in the years to come. The Norwegian Board of Health Supervision estimates that the number of hip replacements done annually will rise by 40% for women and 70% for men by the year 2030 [2]. This highlights the need to better understand how and why implants are failing. In order to do this, clinical researchers need suitable IT tools.

The research question that this development sought to answer was the following - what is the potential impact of transferring the current state of information processing to a database system on the research, information management, and data sharing?

II. METHODS

Hevner, March, Park, and Ram [3] divide information systems research into two distinct paradigms - behavioral science and design science. The former has its roots in the natural sciences and concerns itself with explaining phenomena that explain the development and usage of information systems [4]. Design science on the other hand has its roots in engineering and aims to extend the boundaries of human and organizational capabilities by creating new and innovative artifacts [3]. An important point to make is that the use of the artifacts produced by design science can lead to phenomena that can in turn be studied by behavioral science [4]. This thesis follows the design science research paradigm.

The initial work on user needs assessment began months before the actual system implementation, at a meeting with the users. At this point the limitations of the current solution and high level goals of the potential prototype were discussed. While the technical details of the prototype were still not detailed, the high level architecture plans were clear and the strategy was to implement as a local web application. The reason for this approach was that it might later be further developed into an internet-enabled web application offering access to external parties such as other researchers or clinicians. When the actual project work started a few months later, it became clear that the plans for external access were not in the immediate future. This fact, coupled with security concerns, led to taking a different approach to the prototype architecture. The main prototype would now be built as a desktop application.

Further exploration of the information needs involved semi structured interviews with the users individually. The aim was to allow the researchers to express their thoughts about their information management freely, but still within

H. Mindedal and M. Persson (eds.), *16th Nordic-Baltic Conference on Biomedical Engineering*,
IFMBE Proceedings 48, DOI: 10.1007/978-3-319-12967-9_7

the bounds of the predefined themes. These interviews explored how the users interact with the current system in their daily work. The interview subjects filled different roles in the organization.

The interview sessions concluded with an observation of the users showing examples of their interactions with the system from their workstations. The interviews were very useful at this stage of the project, as they served to increase the understanding of current working practices and processes. Some limitations and ideas for solutions that were not previously discussed also surfaced during these interviews.

A relational database was created to combine the data structures of the locally generated data with the data coming from the national registry.

III. RESULTS

Figure 1 presents an overview of the different parts of the extended system and the information flow between its elements. Elements in green are parts of the developed prototype. The administrative lab user handles importing registry values, while the other users interact with the record data in the desktop application. Potential external users interact with a subset of the same data through a web application.

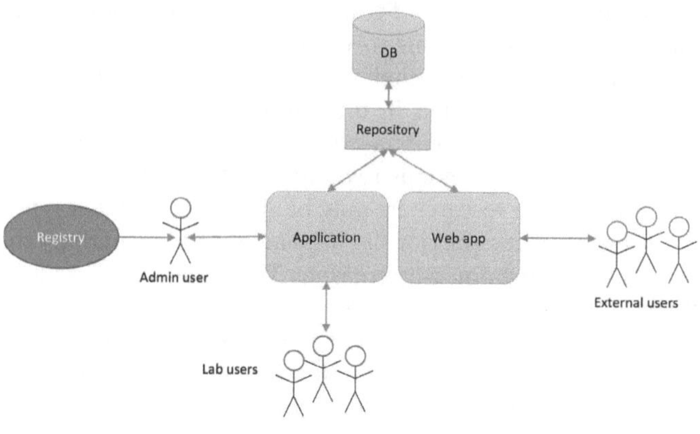

Fig. 1 System overview

Figure 2 shows one of the main screens of the desktop application, where users view and edit information about a specific record. This screen provides the user with an integrated view of both the information entered by the lab users and the information retrieved from the registry, if available. In addition to the desktop application, a basic web application prototype was also developed to demonstrate the feasibility of offering a limited external access to the same database used by the desktop application.

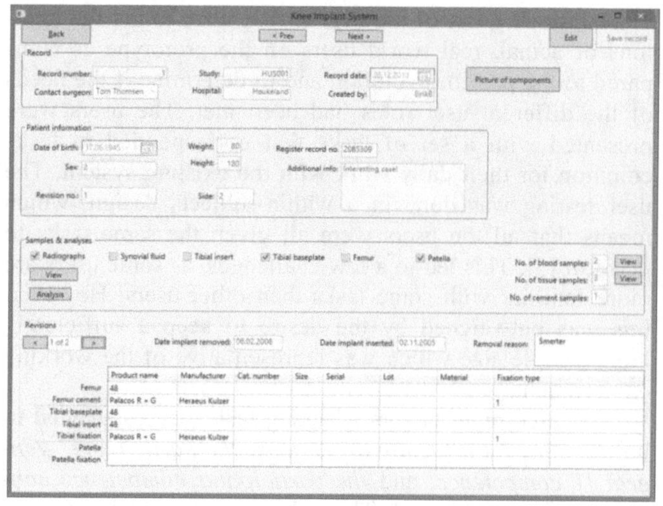

Fig. 2 Detailed record view

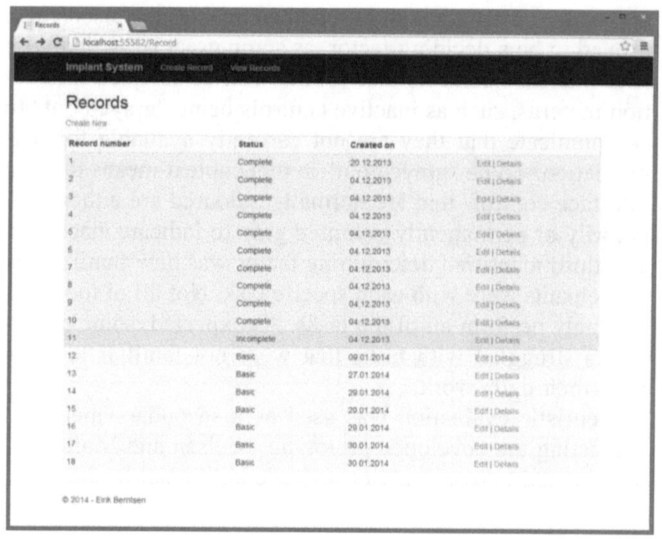

Fig. 3 Web application record list

Figure 3 shows the web application listing all available records.

Two methods of evaluation were used to assess the prototype: user testing and heuristic evaluation. The user testing was carried out with the Haukeland lab staff as participants, while the heuristic testing used master students at the Department of Information Science and Media Studies as participants.

The broad goals of the user testing were to get the opinions of actual, real world users on the prototype as compared to the previous solution and to determine if the needs of the different user roles had been met. The users were presented with a set of tasks that corresponded to tasks common for their daily work with the existing system. The user testing was done in a within-subjects design, which means that all the users were all given the same tasks to perform [5]. This led to a few challenges, as some users are more familiar with some tasks than other users. However, this was outweighed by the desire to keep a sufficiently large sample size which was representative of the working environment.

The three main factors affecting task success seemed to be *the inherent difficulty of the task, the participants' general IT competence*, and *the participants competence concerning the specific task*. The tasks that were the most complex, as defined by the required number of sub-actions needed to complete it, were the tasks with the lowest success rates. The general IT competence of the users also proved to be a deciding factor, as some users struggled with user interface controls that follow well established interaction patterns, such as inactive controls being "grayed out" to communicate that they are not currently available for manipulation. To be "grayed out" in this context means that the interface controls that are normally coloured are either temporarily or permanently coloured gray to indicate inactivity. The third identified determining factor was how familiar the participants were with each specifc task. Not all of the users routinely perform all of the tasks. As expected, some participants struggled with tasks that were not familiar to them from their daily work.

Heuristic evaluation was used as a secondary means of evaluating the developed prototype. Nielsen and Molich [6] recommend to never rely on a single individual for heuristic evaluation and instead use 3-5 participants and aggregate their evaluations.

This approach was followed for the heuristic evaluation of the prototype. Four information science master students were recruited to evaluate the system according to Nielsen's ten heuristics [6]. The evaluators were asked to perform the same tasks as the user testers were, but they would not be timed. They were explained that the session would not assess their performance in any way, but rather focus on potential usability issues. They were also instructed to ask questions if anything was unclear or confusing, especially concerning medical terms in the application that they could not be expected to be familiar with. Whenever they encountered a usability issue, they were asked to press a key combination on the keyboard and verbally describe the issue. These events could then be searched and easily organized afterwards.

Issue no.	2
Location	Create new record
Heuristic	Recognition rather than recall
Issue	There is nothing indicating the correct date format if a user chooses to type in the date instead of using the date picker widget.
Recommendation	The date input fields should have hints suggesting the proper date format (22.04.2013).

Fig. 4 Usability issue

The identified usability issues were scored in terms of severity in order to prioritize the need to address them. The scale suggested by Nielsen [7] was used. It ranges from 0 to 4, '0' being a non-issue and '4' being a catastrophic usability problem. Nine such issues were identified, neither of which were discovered by the user testing.

All of the issues were scored between 1 and 3 in severity. One such issue is summarized in Figure 4. The evaluators also provided their recommendations as how to resolve the identified issues.

IV. DISCUSSION

When developing the prototype, the goal was to implement it while respecting processes that are already in place. The prototype offers two notable improvements as compared to using the networked spreadsheet or Access database to manage the joint implant information. One is the ability to see the history of revisions for a given patient. The other is that users are able to view and edit the records concurrently without experiencing write access restrictions.

One major limitation in the current solution is that the details of the components used in a revision will overwrite the details of any previous revisions for that patient. This is a problem since researchers may be interested in what types of implants were previously implanted, which parts were changed out and for what reason. The prototype remedies this problem by allowing users to browse between the history of revisions and see details about individual components and the reason for removal. In this way the users can get an idea of the patient's implant history.

Excel spreadsheets are not designed to allow concurrent network access from multiple users. If one user has opened the database, the second user who opens it will get a warning that it is locked for editing by the first user. The first user will then have to close the database before the second user can make changes to the data.

The web application prototype will hopefully serve as an example that will enable future developers to easily create web-enabled functions that could interact with this database system. As an example, one such function might be a data mining function to help report creation based on the data contained in database.

The two different evaluation methods used in this thesis served complementary purposes. The user testing examined how the real world users interacted with the system and how well it supported their information flow as compared to the current practices. Performing user testing of the prototype served several purposes. Besides generating feedback and performance metrics, it also served as a demonstration for many of potential users. Prior to the user testing sessions, only some of the users had seen earlier versions of the prototype.

Heuristic evaluation used expert evaluators with more technical insight in order to discover a number of potential usability issues that were not explicitly reported by the user testing. Due to the evaluators having an understanding of the technological standards, limitations and possibilities for such an application, they were also able to offer suggestions on how to improve these identifed issues.

The amount and quality of the feedback and constructive criticism received from both the user testing sessions and the heuristic evaluation have exceeded expectations. Each method provided valuable insights that could be transferred to any potential further development. The results gathered from the evaluation offer concrete suggestions for improvements.

V. CONCLUSIONS

The current prototype has a notable potential to impact the information management at the Biomaterials lab. The application can export records to SPSS format, but additional formats for other types of specialized software could also be added. In this way, the researchers could use the database as a central repository and export the data in an appropriate format for a particular analysis. Using a full-fledged relational database instead of networked spreadsheets, or file based databases such as Access, opens up new possibilities in data sharing, both within the organization and potentially with external parties. Following an established research paradigm such as design science should ensure that the project was carried out in a consistent and predictable fashion so that the traceability of the research remains high.

The developed application has the potential to be further improved to the point of putting it into daily use and supporting researcher with data storage, and administration as well as information retrieval. The prototype demonstrates several improvements as compared to the existing solutions for both the knee and hip implants. Strategies are being developed to include it in the hospital information system (HIS).

ACKNOWLEDGMENT

This project could was carried out with the generous help of the users at the Biomaterials lab of the Haukeland University Hospital in Bergen. The authors are especially thankful to Drs. Peter Ellison and Paul Johan Høl.

REFERENCES

1. Haukeland University Hospital (2013) The Norwegian Athroplasty Register
2. The Norwegian Board of Health Supervision (1999) Scenario 2030
3. Hevner, A. R., March, S. T., Park, J., & Ram, S. (2004) Design science in information systems research. MIS Quarterly, 28(1), 75-105
4. March, S. T., & Smith, G. F. (1995) Design and natural science research on information technology. Decision Support Systems, 15(4), 251-266
5. Rubin, J., & Chisnell, D. (2008) Handbook of Usability Testing: Howto Plan, Design and Conduct Effective Tests. Wiley.
6. Nielsen, J., & Molich, R. (1990) Heuristic evaluation of user interfaces. Proceedings of the sigchi conference on human factors in computing systems empowering people – chi '90 (pp. 249-256). New York, USA. ACM Press
7. Nielsen, J. (1995) Severity Ratings for Usability Problems. Nielsen Norman Group

Author: Eirik Berntsen, M.Sc.
Institute: Department of Information Science and Media Studies, University of Bergen
Street: Postboks 7802, N-5007 Bergen
City: Bergen
Country: Norway
Email: Eirik.Berntsen@student.uib.no

Planning of Maintenance for Medical Devices through a Workload Redistribution in a Clinical Engineering Service

R. Calabrese[1], J. Baldini[2], I. Lasorsa[1], and A. Accardo[1]

[1] Department of Engineering and Architecture, University of Trieste, Trieste (TS), Italy
[2] Centre Operative Managing, Elettronica Bio Medicale s.r.l., Foligno (PG), Italy

Abstract— **The common practice of Clinical Engineering Services (CESs) is to manage separately Electric Safety Checks (ESCs), Quality Controls (QCs) and Preventive Maintenance (PM). However, protocols related to these activities are partially overlapped. This results in a redundancy of maintenance steps producing a waste of time.**

Furthermore, outsourced CESs are in charge of managing activities in hospitals located in different areas with the same team. Thus the distance to be covered by technicians from an area to another one in order to reach the equipment represents a waste of time.

This problem can be solved by merging ESC, QC and PM into a unique operative protocol we call Single Planned Activity (SPA). This implies the development of new maintenance plans and the human resources redistribution, and clinical engineers are in charge of this task.

In order to assess its feasibility, we developed a comprehensive method that consists of the analysis of technician's workload and of the algorithm able to generate a monthly maintenance plan. The method reprograms the previous planned activities through the available human resources redistribution in a more efficient way, according to their actual workload.

The method has been tested in an Italian CES and proved to be useful for clinical engineers.

Keywords— **Clinical Engineering Services, Maintenance Plan, Preventive Maintenance, Workload Analysis.**

I. Introduction

A modern clinical engineer has to face with many diversified tasks depending on functions and priorities of the department in which they work [1].

Moreover, the independent multihospital healthcare technology management systems are growing up, resulting in a medical equipment dislocation on extended areas [2]. So, the maintenance plans are often fragmented and not efficient. Consequently, CESs have to face with time and economic wastes. This happens when temporary needs drive the long-term planning. Indeed, it is very important to provide the clinical engineers with specific tools to make planning activities more efficient.

In order to estimate a priori the size of the staff and its distribution among engineers and technicians, some models have been proposed in the literature [3,4]. In this work we start from an existent CES, and we focus on the effects on the organization caused by changes of processes.

Particularly, we examine the methodology used by an Italian CES for managing tens of thousands medical equipment in different hospitals located in an Italian district. The CES organizes its planned activities involving two different teams: the first one dedicated to the QCs and the PMs, the second one dedicated to the ESCs. These activities are separately managed into three different plans. We present how these different activities can be merged into a SPA and why this solution can be considered more efficient. The solution concerns four main challenges:

1. To train the staff in order to make it able to execute the three abovementioned activities;
2. To merge the different standard operating procedures in a single one;
3. To design and develop a planning algorithm that aims to reduce wastes of time and of economic resources;
4. To assess the staff workload in order to redistribute it into a more efficient way.

This work aims to deal with the points 3 and 4, assuming that points 1 and 2 have been already solved. The algorithm was designed in order to reach the following goals:

1. Minimization of the distance covered by technicians;
2. Balancing the workload among technicians;
3. Saturation of the technicians workdays;
4. Be compliant with recommendations provided by manufactures and international standards about the correct maintenance periodicity for any medical device (MD).

In particular we developed a monthly maintenance plan for technicians in order to allow them to select daily groups of SPAs to be carried out depending on different factors like MD availability or urgent and unexpected corrective maintenance to be done.

In order to assess the technicians workload we analyzed three methods suggested by the literature [5]: the semi-structured interview, the document analysis and the questionnaire. We chose to administer a daily questionnaire to the staff on how they spend their time during the workdays, without interfering with their usual working activities.

H. Mindedal and M. Persson (eds.), *16th Nordic-Baltic Conference on Biomedical Engineering*,
IFMBE Proceedings 48, DOI: 10.1007/978-3-319-12967-9_8

Finally, in order to find the best matching between human resources and tasks to be done, many variables (such as job role, skill level, geographical location, language, potential retraining, etc.) must be considered [6].

II. MATERIALS AND METHODS

A. Workload assessment

The workload assessment was made considering 12 technicians working in two different locations (α, β), 7 in α, and 5 in β. The typical workday of a technician was analyzed through face-to-face interviews with the staff, in order to generate a list of basic tasks they carry out daily. The resulting list aims to be comprehensive of all the possible tasks to be made in a common workday.

In the first column of Table 1 a list of the possible activities of a typical day is presented. Technicians were required to fill the table, with the information related to the number of each specific activity they made, and the time (in min) they spent for each one, in a working day (hour by hour, from 8am to 5pm, as indicated in the header).

They filled one worksheet per day for 10 days. Worksheets were collected in anonymous way, in order to avoid any personal assessment.

The daily records were analyzed in order to get information about the productivity at both α and β locations. Also the average tasks durations were analyzed, but it is not presented in this paper. The daily productivity was calculated as the time spent by technicians in work activities divided by the whole workday duration. The total productivity was calculated as the mean of the daily productivities in the two locations.

B. Algorithm for maintenance schedule

Table 1 Example of a daily worksheet

Task	Number	08:00-09:00	09:00-10:00	…	16:00-17:00
Receiving request for intervention	…	…	…	…	…
Internal Repair	…	…	…	…	…
Updating information system	…	…	…	…	…
…	…	…	…	…	…

The designed algorithm (developed in Matlab® environment) needs two inputs: the first concerns the available labor force, in terms of number of technicians and related to the part of their workday that is expected they had to spend in SPAs during the year; the second input is represented by the previous annual maintenance plan that consists of an inventory in which MDs are listed. Each row, corresponding to one MD, includes information about its location, maintenance history and the expiration dates for ESC, QC and PM.

In the first step of the algorithm, data are picked up from the previous maintenance plan and transferred to a purpose-built table. Two new data are added to each row:

1. The date of next expected SPA, according to the periodicity assumed as the smallest among ESC, QC and PM;
2. The estimated durations of ESC, QC and PM, assessed through the user manual indications of the DM and interviews with technicians.

The table is sorted by the date of next expected SPA in ascending order.

Two variables were used: the first represents the time available in a workday (daily-buffer), while the second one is a vector with 12 elements representing the time available in each months of a year (monthly-buffer). We set the daily-buffer to 8 hours (a normal workday) and the monthly-buffer depending on the holydays and force labor expected.

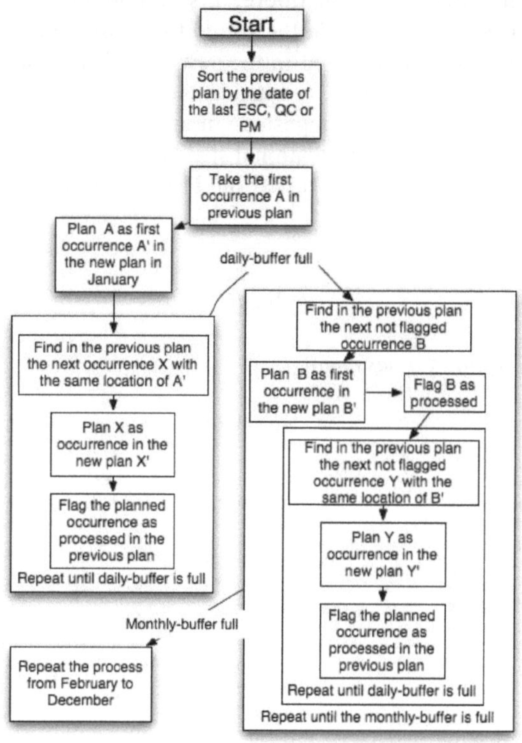

Fig. 1 Conceptual scheme of the algorithm

In Figure 1, it is represented a conceptual scheme of the algorithm.

The algorithm takes the first occurrence in the previous plan, adds the estimated duration of the latter, filling the information in both the daily and the monthly buffers, starting from January. This results in the first planned SPA in the new plan, and all the needed information (e.g., ID number of the MD and its locations) are saved. The total length of the SPA is calculated as the sum of the durations of the three activities. Once the SPA has been planned for the next year, it is flagged as 'processed' in the sorted table. The first planned activity location drives the subsequent planned activities. Indeed, the algorithm proceeds in the sorted table analysis searching and selecting the planned activities with the same location and adding them into the final planning table until the daily-buffer is full. Consequently, the algorithm restarts in the sorted table analysis from the top, assumes as driver the first occurrence not flagged and repeats the described process until the monthly-buffer for January is full. So, algorithm repeats continuously this operation until all the occurrences in the sorted table are processed.

This approach allows saturating the workdays, thanks to the daily-buffer, and respects the expected force labor per month, thanks to the monthly-buffer. Taking as driver for the planning the MD location, the algorithm optimizes technicians' travels. Furthermore, it's important to note that occurrences expected in the sorted table can be only anticipated and not postponed in the new plan. This allows algorithm to avoid delays in making the planned activities according to the suggested minimum frequency by manufactures and international standards.

The algorithm was tested on 1191 MDs divided into 3 areas (x, y, z).

III. Results

A. Workload assessment

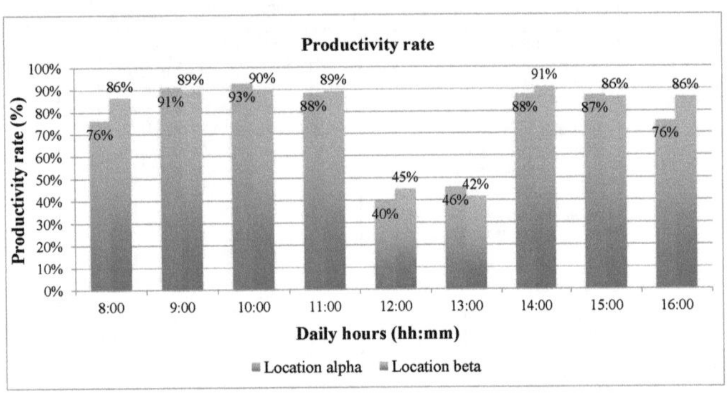

Fig. 2 Location α: Productivity

Workload was estimated for the two groups of technicians operating at the α and β places. 70 sampling sheets was expected for α and 50 for β, but due to the occasional absences from work of technicians, they resulted in 68 sampling sheets for the location α and 50 sampling sheets for the location β.We extracted from worksheets the productivity rate for each hour in a day, in order to assess its daily trend (Figure 2). The 100% of productivity means that all the time available for technicians in a workday was spent in work activities among those listed in each worksheet. No difference between the two locations is observed; at the lunch time, as expected, productivity rate decreases due to the lunch break.

B. Algorithm for maintenance schedule

The application of the algorithm produces the final plan consisting of a monthly list of SPAs to be done within a year, grouped into packages of duration not longer than a workday. Moreover, thanks to the MD-location used as driver in order to saturate a daily-buffer, travels among the different hospitals are reduced.

The typical scenario is that a technician will be provided with the monthly plan within the grouped SPAs to be done. So, he can move to the first location and carry out the SPAs planned. The algorithm builds these groups of SPAs in order to saturate their workday and to reduce their travels among hospitals (one travel maximum a day).

The algorithm results in a final table of the planned activities during a year. It is possible to evaluate the total amount in hours of the planned workload for each month of a year and compare it with the force labor expected at the beginning.

Figures 3, 4 and 5 show the results representing the force labor expected (blue lines) and the workload resulted from the application of the algorithm (red lines) in the three considered areas (x, y, z).

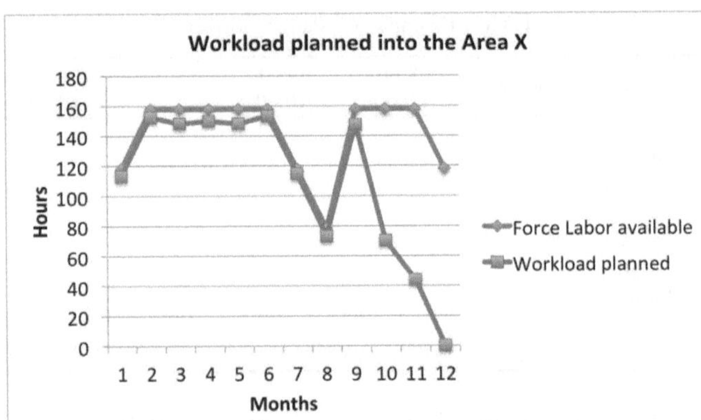

Fig. 3 Summarized results by applying algorithm into the area X

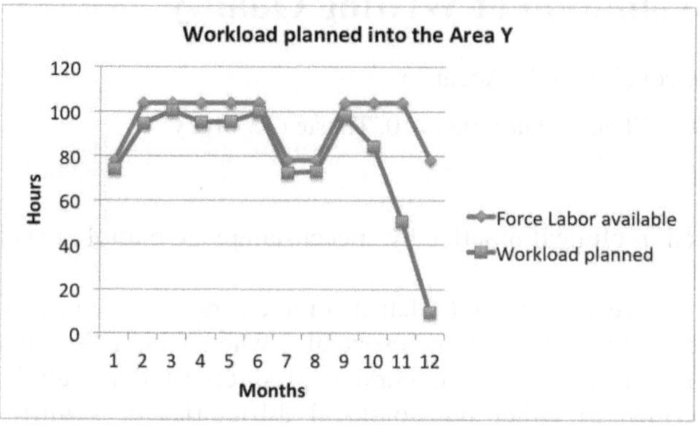

Fig. 4 Summarized results by applying algorithm into the area Y

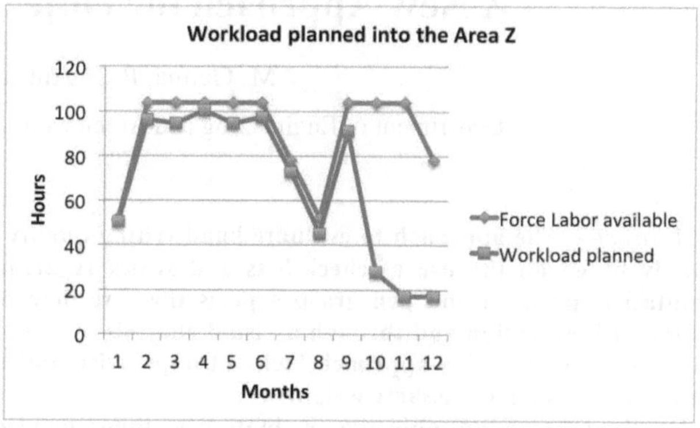

Fig. 5 Summarized results by applying algorithm into the area Z

IV. Discussion

A. Workload assessment

The productivity rate (Figures 2) is always very high during the day, in both the locations with an expected inflection during the launch break. In fact, in these two hours the staff alternates for about one hour each.

Slight decreases are positioned both at the start and at the end of the workday, while, during the middle of the day, the productivity rate is stable.

B. Algorithm for maintenance schedule

The Figures 3, 4 and 5 show as the algorithm is perfectly capable to fill the maintenance plan with the force labor expected. In all the cases, the workload planned fills almost perfectly the force labor expected until there are not further occurrences to be done. At the ninth month, in all the three graphs, workload planned curve stops to fill the force labor. The gap between the two curves represents that force labor in the last months is overestimated compared to the load.

V. Conclusions

In this work we applied an optimization algorithm based on a CES. The results proved the method to be an effective tool for planning an efficient workload.

The workload assessment promises to provide clinical engineers a tool for restructure the human resources distribution according to the workload expected. In the future, the method should be tested on larger inventories and more complex CESs.

Acknowledgment

Work partially supported by the University of Trieste, Master in Clinical Engineering.

References

1. Frize M (1990) Results of an international survey of clinical engineering departments. Medical and Biological Engineering and Computing 28(2):153-159
2. Clark J (2004) Challenges facing independent multihospital healthcare technology management systems. Engineering in Medicine and Biology Magazine IEEE 23(3):20-26
3. Balestra G, Gaetano L, Puppato, D (2008) A model for simulation of Clinical Engineering Department activities. Engineering in Medicine and Biology Society 30th Annual International Conference of the IEEE pp 5109-5112
4. Lamberti C, Panfili A, Gnudi G et al. (1997) A new model to estimate the appropriate staff for a clinical engineering department. Journal of clinical engineering 22(5):335-341
5. Souza C, Jerico M, Perroca M (2014) Measurement of nurses' workload in an oncology outpatient clinic. Revista da Escola de Enfermagem da USP 48(1):97-103
6. Naveh Y, Richter Y, Altshuler Y, et al. (2007) Workforce optimization: Identification and assignment of professional workers using constraint programming. IBM Journal of Research and Development 51(3.4):263-279

Author: Raffaele Calabrese
Institute: University of Trieste
Street: Via Valerio, 10
City: Trieste
Country: Italy
Email: raffaele.calabrese@phd.unit.it

A New Approach for Objective Evaluation of Writing Quality

M. Genna, P. D'Antrassi, M. Ajčević, and A. Accardo

Department of Engineering and Architecture, University of Trieste, via Valerio 10, Trieste (TS), Italy

Abstract— **The approach to evaluate handwriting quality is usually based on the use of check-lists and scales regarding qualitative, postural and pen grasp aspects that are detected during task execution and through a visual analysis of written texts. Nevertheless, this approach lacks of objectivity and all the parameters have the same weight.**

On the other hand, objective evaluation methods focusing on the qualitative approach are available in the literature. For example, the analytic hierarchy process (AHP) is a technique that allows to organise complex relationships between elements into a system based on subjective judgements.

In this paper we present an innovative evaluation scale, based on a new neuromotor model of handwriting production, which associates a specific weight to each parameter during the calculation of a quality score. The scale allows teachers, speech therapists and all those interested in diagnostic and rehabilitative phase to identify subjects who show some handwriting difficulties compared to the average performance of their peers.

The new approach was applied to two groups of students, following different teaching methods, in order to evaluate suitably weighted aspects of handwriting quality.

Keywords— **handwriting, quality evaluation, dysgraphy, teaching methods.**

I. INTRODUCTION

The qualitative evaluation of handwriting goodness is usually based on the use of check-lists and scales regarding postural and pen grasp aspects, which are observed during task execution, and visual analysis of the written product. Most of the handwriting evaluation scales, developed during the last 35 years, are based on criteria of *readability* highlighting the problem of objectivity and the need for standard [1]. Handwriting quality measurement, especially of cursive writing, is not as easy as it could seem, since it needs to consider not only the graphomotor sequence of each grapheme but also its space arrangement.

Nowadays, writing evaluation is based on empirical evidence and subjective experience. Therefore, within clinical practice, new methods scientifically validated and as objective as possible are necessary.

A new approach for the evaluation of effector components involved in the realization process of cursive writing is proposed. It uses both validated evaluation scales, most commonly used in the literature [2], and observational grids, used in clinical practice by speech-language pathologists [3, 4].

It's very common to detect some errors or inaccuracies in the writing of children (especially when facing the movements required in the cursive production) that could be ascribed to either physiological difficulties or significant handwriting disorders. The discrimination between writing attitude delay and dysgraphy can be realized by basic skills (i.e., personal and extrapersonal spaces organization) evaluation [5]. One of the latest tools that solve this problem is the test for the evaluation of graphomotor and postural difficulty DGM-P [6] which compares the writing performance level of children marked as slow and illegible handwriting with the typical performance of their peers with experience in the writing learning. However, this approach lacks of objectivity and all the parameters have the same weight.

The aim of this work is to overcome these shortcomings by developing a new approach for objectively evaluating writing quality. A first application of this tool to compare the traditional method for teaching writing with the alternative space-time method of Ida Terzi [7] is presented.

II. MATERIALS AND METHODS

A. Participants

The novel tool was tested on two samples of about 20 students recruited from the "Don Milani" primary school of Cernusco sul Naviglio, that follows the Ida Terzi's method (group labeled with CE), and the primary school of Pioltello, that instead uses the traditional teaching method (control group, labeled with PI). The acquisitions were made at the end of 2th grade of Primary school, the first year of learning cursive handwriting. All subjects were Italian mothertongue, right-handed and without handwriting problems and organic pathologies.

B. Teaching methods

Ida Terzi, primary school teacher at the institute for the blind in Reggio Emilia, Italy, in the first half of the 1900s proposed a space-time method with the aim to develop students' perception of the body moving in space so that they could integrate information from personal (own body), peri-personal (objects manipulation) and extra-personal

H. Mindedal and M. Persson (eds.), *16th Nordic-Baltic Conference on Biomedical Engineering*,
IFMBE Proceedings 48, DOI: 10.1007/978-3-319-12967-9_9

(environment) space facilitating perceptual consistency and transition from using an unconscious to a conscious use of the body in motion. This technique is supported by different theories of neuroscience that underline the importance of the motor system as a tool for knowledge: embodied cognition (role of the body in the formation of cognitive and social processes) [8], constructivism (knowledge of reality is the result of a personal elaboration based on experiences) [9] and the theory of mirror neurons (neurons are activated not only when an action is performed, but also when movements made by others are observed) [10].

C. Tests

All participants wrote in cursive, first in accurate (A test) and then in fast (V test) mode, the Italian sentence: *in pochi giorni il bruco diventò una bellissima farfalla che svolazzava sui prati in cerca di margherite e qualche quadrifoglio.*

This phrase was constructed to containing all the letters of the Italian alphabet and several phonological rules.

D. Neuromotor Model

The implemented neuromotor model of handwriting generation integrates the two-way model [11], with the space-time method proposed by Ida Terzi (Figure 1).

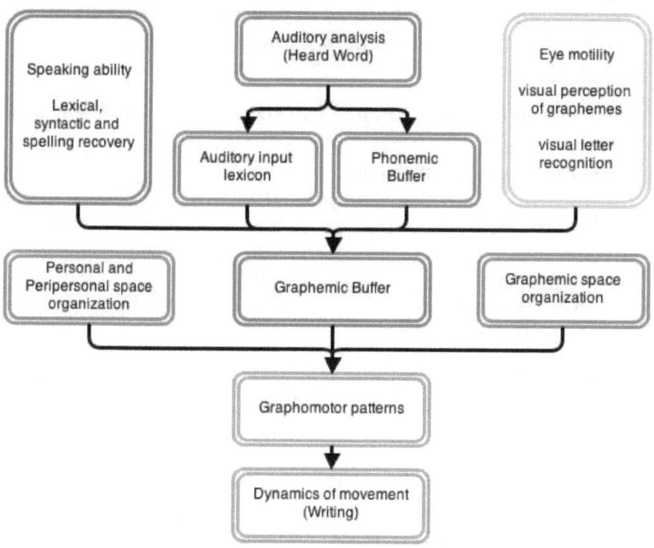

Figure 1 Neuromotor model of handwriting generation

In the model the writing is the result of: a) speaking ability, involved in the process of spontaneous production of a text; b) analysis and identification of the phoneme, activated by writing from dictation; c) visual perception, triggered by coping of a text; d) graphemic buffering, i.e. the recovery of the grapheme to reproduce; e) grapho-motor pattern generation, which identifies the sequence of movements necessary for the realization of the grapheme; and f) movement dynamics, in which the efferent components are activated. All steps are guaranteed by a prior personal, peri-personal and extra-personal space organization.

E. Evaluation Scale

The evaluation scale considers the terminal part of the model, focusing on posture and pen-grip definitions as well as on spatial placement of the written elements and their sequence and dynamics. Each model's area (*personal and peri-personal spaces organization, extra-personal space organization, graphemic buffer, graphomotor patterns*) is characterized by several parameters. To each parameter is associated a normalized score, calculated by the ratio between the number of errors done and the maximum number of errors that can be done in the test. For example, the parameter *variability of the alignment left* is the ratio between the number of rows in which the alignment is not respected by the student and the total number of rows. Thus, each dimensionless score assumes a value between 0 (no error) and 1 (maximum number of errors).

These scores allow to perform the assessment on the error rate compared to what is actually written, contrary to what is provided by the current rating scales that use absolute and not normalized scores. This represents a novelty introduced by the implementation of the proposed evaluation process.

Table 1 Evaluation scale

Model Area	Sub Area	Parameter/deficit/criteria
Personal and Peri-personal spaces organization	posture	inefficient posture
	handgrip	inefficient handgrip
Extra-personal space organization	sheet graphic space	variability of the alignment left
		right margin is not respected
		irregular line spacing
	row graphic space	irregular word spacing
		letters collision
		fluctuations on the line
		max variation of letter size
		wrong letter size
Graphemic buffer	recovery allographic	confusion between visuo-spatial similar letters
		confusion about the type of character
Graphomotor patterns		wrong graphomotor pattern
		dysmetria in the execution of the trajectory of the letter
		incorrect binding between letters
		self-corrections of grapheme written

F. Total Quality Score

Another innovative element of the implemented assessment tool consists of the association of a specific weight to each parameter in the Total Quality Score (TQS) calculation. TQS, given by the sum of the weighted contributions of individual parameters, will allow the comparison of the handwriting quality between subjects; helping teachers, speech therapists and all those interested in the diagnostic-rehabilitative phase in the identification of subjects presenting some difficulties compared to the average performance of the group.

In order to determine the relative weights of the parameters (so identifying the parameter that most affects the handwriting quality) the AHP method [12] is used.

The AHP method allows to assign a priority to a number of decisional alternatives and to relate parameters characterized by qualitative and quantitative assessments, although not directly comparable, by combining multi-dimensional measures into a single priorities scale. The AHP method steps implemented in our work are:

1. identification of priorities through pairwise comparison of each parameters of the evaluation scale in order to find which one affects more the writing quality;
2. consistency check of judgment in all pairwise comparisons through the use of the *Super Decisions* SW [13]

G. Processing and statistical analysis

In order to investigate the influence of each teaching method on the writing quality, for both groups and in each test, mean value and standard deviation of: TQS, each model's sub-area and each parameter (both listed in Table 1) were calculated. For each of test, the significance of the difference between the two groups using the Wilcoxon test for independent samples was evaluated.

In order to identify the most significant parameters for each test, stepwise regression with forward selection was used. Furthermore, each variable was included in a multilinear model only if it was statistically significant ($p<0.05$).

Moreover, to assess possible differences between the control class and the Cernusco's class considered, principal component analysis (PCA) was performed on the normalized (Z- score) parameters selected by stepwise regression.

III. Results

Starting from TQS, a significant difference ($p < 0.0001$) between the two groups is detected, in particular the Cernusco's class has a value of 0.08 ± 0.02 while the control group a value of 0.14 ± 0.03.

In detail, it is possible analyze the single model sub-areas to discover in which the two schools differ. Table 2 shows that, in the A test, for all sub-areas the Pioltello's group totalizes higher scores than the other school, thus producing more errors.

Table 2 Mean TQS by sub-area for A test (*: p<0.005)

Sub Area	Total Quality Score		
	Max	PI	CE
Posture	0,031	0,007	0,004
Handgrip	0,038	0,025	0,023
Sheet graphic space	0,067	0,007*	0,002
Row graphic space	0,328	0,053*	0,031
Recovery allographic	0,169	0,001	0,001
Graphomotor patterns	0,366	0,052*	0,023

Tables 3 to 5 show in detail the criteria for the areas (Table 2) in which significant differences are observed (*).

Table 3 Mean TQS by *sheet graphic space* criteria for A test (*: p<0.005)

"Sheet graphic space" criteria	Total Quality Score		
	Max	PI	CE
variability of the alignment left	0,020	0,001	0,001
right margin is not respected	0,019	0,006*	0,002
irregular line spacing	0,028	0,000	0,000

About the *sheet graphic space* area, the differences are significant only for the *right margin is not respected.*

Table 4 Mean TQS by *row graphic space* criteria for A test (*: p<0.005)

"Row graphic space" criteria	Total Quality Score		
	Max	PI	CE
irregular word spacing	0,032	0,005	0,002
letters collision	0,122	0,001	0,000
fluctuations on the line	0,052	0,003	0,001
max variation of letter size	0,047	0,021*	0,016
wrong letter size	0,075	0,023*	0,012

The significant differences in *row graphic space* area are related to: *max variation of letter size* and *wrong letter size.*

Table 5 Mean TQS by graphomotor patterns criteria for A test (*: p<0.005)

"Graphomotor patterns" criteria	Total Quality Score		
	Max	PI	CE
wrong graphomotor pattern	0,118	0,022*	0,007
dysmetria in the execution of the trajectory of the letter	0,085	0,013*	0,007
incorrect binding between letters	0,092	0,016*	0,008
self-corrections of grapheme written	0,071	0,002	0,002

For the *graphomotor patterns* area, almost all the criteria are significantly greater in the PI group that in CE one.

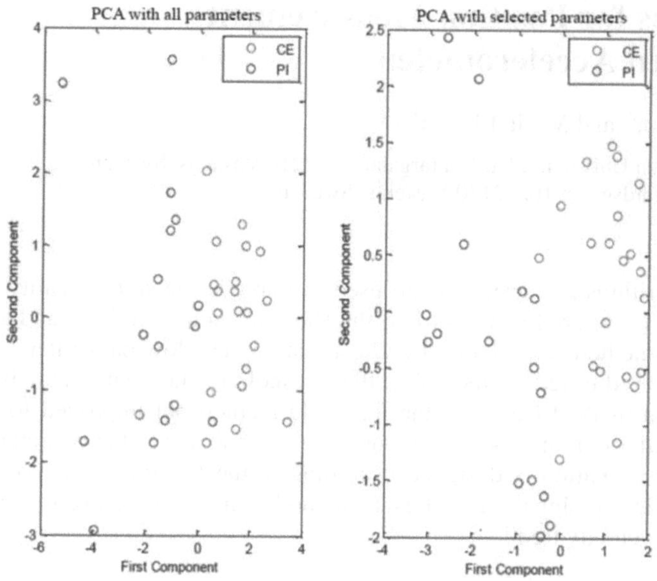

Figure 2 Loading PCA plot for A test

In Figure 2 the loading PCA plots, in the A test, performed using either all quality parameters (left panel) or only parameters selected by stepwise regression (right panel) are shown. The six considered parameters were: inefficient posture, right margin is not respected, wrong letter size, confusion between similar letters, wrong graphomotor pattern and dysmetria. The first two PCA components have a variance associated of 37% (left) and of 57% (right), respectively. A very clear separation between the two groups is observed. Similar results are obtained for the V test.

IV. Discussion

The Terzi's method leads to write with movements as continuous as possible taking larger duration time in writing as well as shorter pen lift duration, as seen in previous studies based on kinematics analysis [14]. This fact could explain the better qualitative performance (lower TQS) by children of the CE group compared to the control PI group, corresponding to a written product of better quality by Cernusco's pupils.

The main differences related to the model's areas (*sheet graphic space*, *row graphic space*, *graphomotor patterns*) are in accordance with the space-time method peculiarity: focalization on improving personal, peri-personal and extra-personal space organization.

V. Conclusion

The new approach described in this paper can be used in the evaluation of teaching methods and help to highlight possible problems which require closer attention for early intervening to prevent dysgraphia.

Acknowledgment

Work partially supported by the University of Trieste, Master in Clinical Engineering. The authors thank Dr.Iolanda Perrone, speech therapist at ULSS 7 Pieve of Soligo, Paola Ceschia and Carmela Mandarino, primary school teachers at Don Milani of Cernusco sul Naviglio.

References

1. S.Rosenblum, P.L.Weiss, S.Parush (2003) Product and Process Evaluation of Handwriting Difficulties: A review. Educational Psychology Review 15(1): 41-81
2. L.Hamstra-Bletz, J.DeBie and B.Den Brinker (1987) Concise evaluation scale for children's handwriting. Swets & Zeitlinger, Lisse
3. P.Tressoldi and C.Cornoldi (2000) Batteria per la valutazione della Scrittura e della Competenza Ortografica nella Scuola dell'Obbligo. Giunti OS, Firenze
4. P.Giovanardi Rossi and T.Malaguti (1994) Valutazione delle abilità di scrittura. Analisi dei livelli di apprendimento e dei disturbi specifici. Manuale generale per la Scuola elementare. Erickson, Trento
5. M.Prateli (1995) Disgrafia e recupero delle difficoltà grafo-motorie. Erickson, Trento.
6. M.Borean, G.Paciulli, L.Bravar, S.Zoia (2012) Test DGM-P - Test per la valutazione delle difficoltà grafo-motorie e posturali della scrittura. Erickson, Trento.
7. I.Terzi (1995) Il Metodo spazio-temporale, basi teoriche e guida agli esercizi. Ghedini, Milano.
8. J.M.Iverson & E.Thelen (1999) Hand, mouth and brain. The dynamic emergence of speech and gesture. J Consciousness Stud, 6:11-12.
9. S.Sjøberg (2007) Constructivism and learning. International Encyclopaedia of Education. 2010, pp 485-490. Elsevier, Oxford.
10. G.Rizzolatti & L.Vozza (2008) Nella mente degli altri. Neuroni specchio e comportamento sociale Zanichelli, Bologna.
11. G.Denes & L.Cipollotti (1990) Manuale di neuropsichiatria: Dislessie e disgrafie acquisite, pp 423-463. Zanichelli, Bologna.
12. T.L.Saaty (1980) The Analytic Hierarchy Process. McGraw-Hill, NY.
13. R.W.Saaty (2003) Decision making in complex environments. At http://www.superdecisions.com/.
14. A.Accardo, M.Genna, I.Perrone, P.Ceschia, C.Mandarino (2014) Evaluation of different handwriting teaching methods by kinematic analysis. XIII MEDICON 2013. IFMBE Proceedings 41:579-582.

Address of the corresponding author:

Author:	D'Antrassi Pierluigi
Institute:	Department of Engineering and Architecture, University of Trieste
Street:	Via Valerio, 10
City:	Trieste (TS)
Country:	Italy
Email:	pierluigi.d'antrassi@phd.units.it

Signal Processing Algorithms for Position Measurement with MEMS-Based Accelerometer

J. Du[1,2], C. Gerdtman[2], and Maria Lindén[1]

[1] School of Innovation, Design and Engineering, Mälardalen University, Gurksaltargatan 9, 72218 Västerås, Sweden
[2] Motion Control i Västerås AB, Ängsgärdsgatan 10, 72130 Västerås, Sweden

Abstract— **This paper presents signal processing algorithms for position measurements with MEMS-accelerometers in a motion analysis system. The motion analysis system is intended to analyze the human motion with MEMS-based-sensors which is a part of embedded sensor systems for health. MEMS-accelerometers can be used to measure acceleration and theoretically the velocity and position can be derived from the integration of acceleration. However, there normally is drift in the measured acceleration, which is enlarged under integration. In this paper, the signal processing algorithms are used to minimize the drift during integration by MEMS-based accelerometer. The simulation results show that the proposed algorithms improved the results a lot. The algorithm reduced the drift in one minute by about 20 meters in the simulation. It can be seen as a reference of signal processing for the motion analysis system with MEMS-based accelerometer in the future work.**

Keywords— **Accelerometer, integration, MEMS, signal processing, drift.**

I. INTRODUCTION

The motion analysis system is an embedded sensor system for health application, human motion analysis and rehabilitation support, which is intended to evaluate a person's ability to move, for example, the neck, arm or leg with the aim of training or rehabilitation. The system [1] is developed with MEMS-based sensors fixated by different wearable bands such as headbands, armbands, and leg bands. The MEMS-based sensors are used for movement measurements and especially position determination and movement patterns measurement.

MEMS-based accelerometer measures the acceleration directly. By integration, the velocity and position can be derived from the measured acceleration. However, measured acceleration is usually interfered with errors such as noise, offset and drift, which are enlarged under integration. So simple and effective signal processing algorithms are desired to solve the problem, improve the result and guarantee the accuracy of the measurement system.

The aim of this paper is to design and evaluate signal processing algorithms for the reduction of the drift of MEMS-based accelerometer. Several methods and algorithms are designed and used to improve the signal quality. These are low-pass filter, threshold method with delay and a method for velocity trend estimation. The low-pass filter is used to reduce the noise; the threshold method with delay is a method based on the threshold method but improved for the continuous movements; the method for velocity trend estimation is designed according to the features of the velocity signals and it is a method which can estimate the velocity trend.

II. METHODS

The signal processing for the position measurements with MEMS-based accelerometers includes integrations and filtering of integrated signals. The block diagram can be seen on Fig. 1.

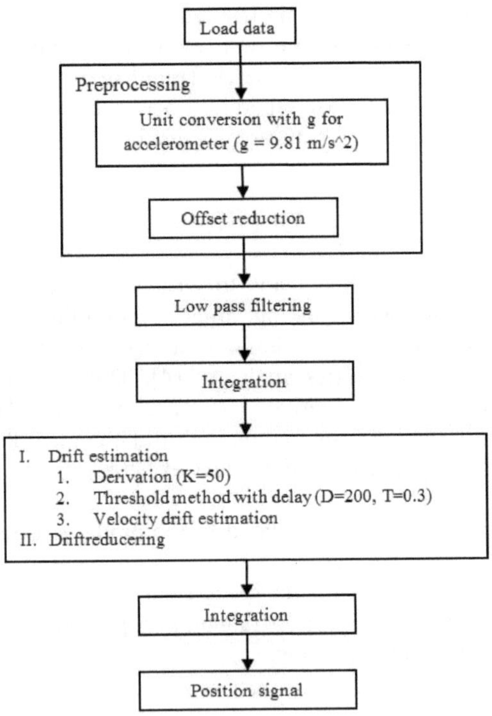

Fig. 1 Flow diagram of signal processing for position measurement with MEMS-based accelerometer.

H. Mindedal and M. Persson (eds.), *16th Nordic-Baltic Conference on Biomedical Engineering*,
IFMBE Proceedings 48, DOI: 10.1007/978-3-319-12967-9_10

As shown in Fig.1, after preprocessing of measured acceleration signal, the acceleration signal is filtered by low-pass filter in order to reduce noise and integrated to velocity. Preprocessing usually contains the reduction of noise, offset and temperature drift [2]. The velocity signal after integration has drift which can lead to more errors in the integrated position signal. So methods with threshold method with delay and velocity drift estimation method are designed and used to estimate the velocity drift. After reduction of the velocity drift, the position signal is obtained directly from the integration of filtered velocity signal.

The algorithms used are introduced as follows.

A. Integration

For the accelerometer signal, the position can be got by two integration of the measured acceleration signal.

$$s = \int \vec{v}dt = \int (\int \vec{a}dt)dt \quad (1)$$

Because of the digital data and error in the integration of digital data, the Trapezoidal method [3] is used to get a more accurate result of integration.

For accelerometer, the integration calculation is:

$$\theta_{samplen} = \theta_{samplen-1} + \left(\dot{\theta}_{samplen-1} + \frac{\left(\dot{\theta}_{samplen} - \dot{\theta}_{samplen-1}\right)}{2} \right) \times T \quad (2)$$

$$s_{samplen} = s_{samplen-1} + \left(v_{samplen-1} + \frac{\left(v_{samplen} - v_{samplen-1}\right)}{2} \right) \times T \quad (3)$$

B. Filtering

a) Low-pass filter:

A low-pass filter has its cutoff frequency adapted to let the appropriate frequencies below pass. The cutoff frequency can be described as a parameter with a value from zero to one.

A low-pass filter can be described as [4]:

$$y_i = y_{i-1} + \alpha\ (x_i - y_{i-1}) \quad (4)$$

(4) α is the parameter of cutoff frequency for the high-pass filter. x_i is input signals at time point i; y_i and y_{i-1} are output signals at time point i and $i-1$ respectively.

b) Threshold method with delay:

Threshold method with delay is based on the simple threshold method in which the value is set to zero if the value is smaller than threshold. However, the drawback of the threshold method is that there is often not a smooth transition at the threshold setting point and the signal becomes choppy, ex. suddenly becomes zero. Thus, the threshold method with delay is designed to improve the signal under threshold such as continuous movement signal. As shown in Fig. 2, the algorithm can be implemented by the following steps:

1. First is the comparison input signals with threshold parameter. If the signal is larger than threshold at time point $i-1$ and smaller than threshold from time point i , the time calculation is set to zero and output signal is the input signal without processing. If at the next time point the signal is still smaller than threshold, the time calculation adds.
2. Then compare the calculated time with delay parameter, if the input signal value is smaller than threshold and calculated time value is larger than delay parameter, the signal is set to zero; otherwise, the output is just input without processing.
3. Go back to point 1 for next value vid time point $i+1$.

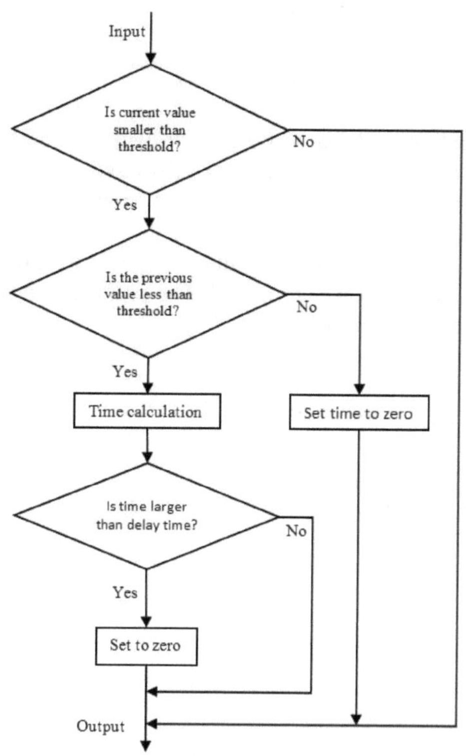

Fig. 2 Flowchart of threshold method with delay.

Threshold method with delay can be described as:

$$y_i = \begin{cases} x_i, x_i \geq T \parallel t_{x<T} < D \\ 0, x_i < T \,\&\, t_{x<T} > D \end{cases} \quad (5)$$

(5) T is the parameter of threshold and D is the parameter of delay. x_i is input signal at time point i ; y_i is output signal at time point i . $t_{x<T}$ is the accumulated time while input signal is small than threshold value.

c) Method for velocity drifts estimation:

Velocity drift estimation method is designed according to the features of velocity signal which is similar to sinus signal.

The velocity drift estimation method can be described as:

$$y_i = \begin{cases} x_i, A_i = 0 \\ y_{i-1}, A_i \sim= 0 \end{cases} \quad (6)$$

(6) x_i is input signal at time point i ; y_{i-1} is output signal at time point $i-1$. A_i is derivation of input signal which can be got from:

$$A_i = K * abs(x_i - x_{i-1}) \quad (7)$$

(7) K is a scale parameter.

III. Results

Simulation is done with raw data which is measured acceleration signal from MEMS-based accelerometer ADXL345 [5] without tilt. The measurement is done with several up-and-down movements.

After preprocessing, the unit of acceleration signal is converted and offset is removed. The preprocessed acceleration signal is shown in Fig. 3.

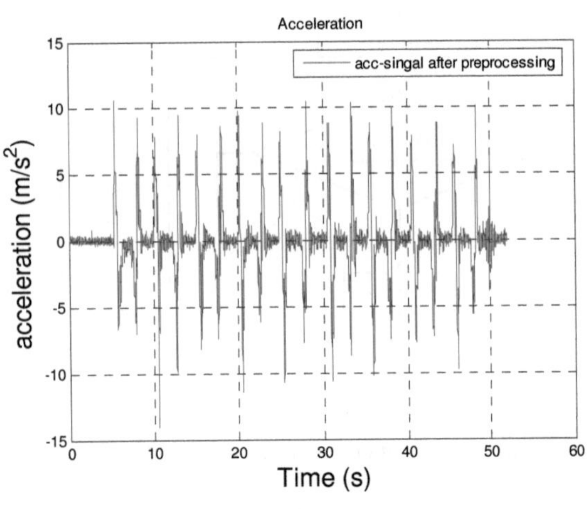

Fig. 3 Acceleration signal after preprocessing.

The velocity signal is got from integration of acceleration signal after preprocessing and low-pass filtering. The integrated velocity signal and its estimated drift are shown in Fig. 4.

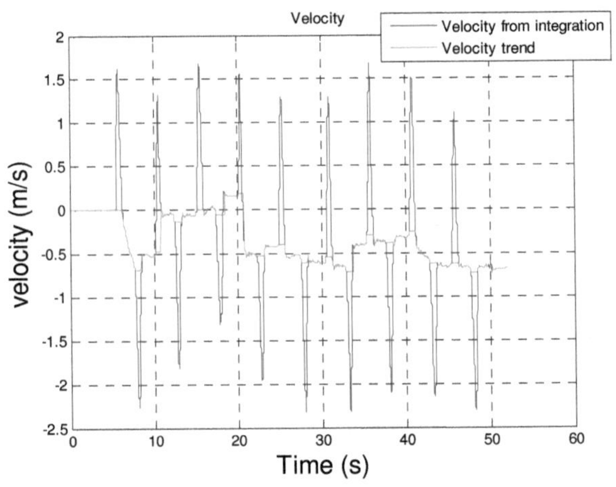

Fig. 4 Integrated velocity signal and estimated drift.

Velocity without drift is integrated to position signal. As shown in Fig. 5, compared with the position without filtering, the drift in integrated position signal is reduced and the result is improved a lot, about 20 meter drift in one minute is reduced.

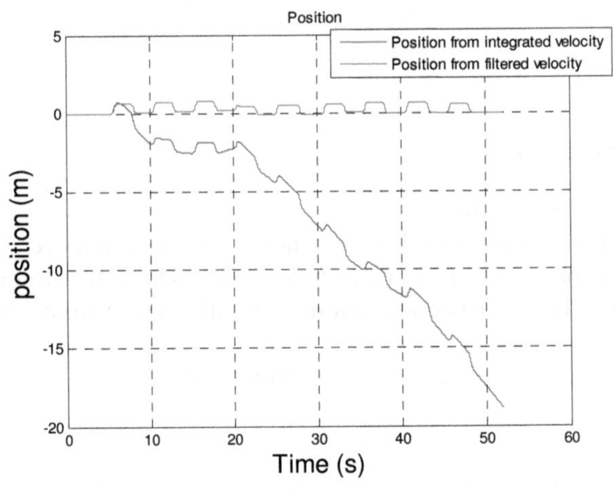

Fig. 5 Position signal comparison.

In summary, the acceleration signal after preprocessing, the integrated velocity signal after drift reduction and the integrated position signal results are shown together, which can be seen in Fig. 6.

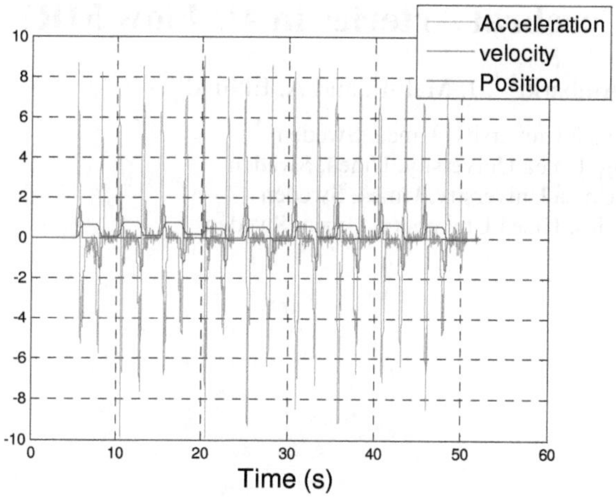

Fig. 6 Acceleration, velocity and position signals.

The results of simulation show that the algorithms used for the reduction of drift especially the integrated velocity drift give a much better result.

IV. CONCLUSIONS

Simple algorithms which are low-pass filtering, threshold method with delay and velocity drift estimation method are designed and used to reduce the drift from integration. Simulation results show that the designed methods reduce the velocity drift and thus the position result is improved a lot. The algorithm reduced the drift in one minute by about 20 meters. These methods are simply designed because of the limited calculation capacity of microprocessor. Based on these developed methods, signal processing for the motion analysis system with MEMS-based gyroscope and accelerometer is expected in the future work.

REFERENCES

[1] Datasheet of TDG system at http://motioncontrol.se/images/tdg-material/Faktablad%20TDG.pdf

[2] J. Du, C. Gerdtman, N. Grbic and S. Johansson. Examensarbetet i utveckling och undersökning av algoritmer för MEMS-gyron, 30hp, avancerad nivå. September, 2012.

[3] Seifert K and Camacho O (2007) Implementing Positioning Algorithms Using Accelerometers. Application Note. Freescale Semiconductor. AN3397

[4] Low-pass filter at https://www.princeton.edu/~achaney/tmve/wiki100k/docs/Low-pass_filter.html

[5] Datasheet ADLX345 at http://www.analog.com/static/imported-files/data_sheets/ADXL345.pdf

Author: Jiaying Du
Institute: School of Innovation, Design and Engineering, Mälardalen University
Street: Gurksaltargatan 9
City: Västerås
Country: Sweden
Email: jiaying.du@mdh.se

Towards Automatic Identification of Cerebral Arteries in 4D Flow MRI

T. Dunås[1], A. Wåhlin[1,2], L. Zarrinkoob[3], K. Ambarki[1,4], J. Malm[3], and A. Eklund[1,4]

[1] Department of Radiation Sciences, Umeå University, Umeå, Sweden
[2] Umeå Center for Functional Brain Imaging, Umeå University, Umeå, Sweden
[3] Department of Clinical Neuroscience, Umeå University, Umeå, Sweden
[4] Centre for Biomedical Engineering and Physics, Umeå University, Umeå, Sweden

Abstract— **4D flow MRI is a powerful imaging technique which provides an angiographic image with information about blood flow in a large volume, time resolved over the cardiac cycle, in a short imaging time. This study aims to develop an automatic method for identification of cerebral arteries. The proposed method is based on an atlas of twelve arteries, developed from 4D flow MRI of 25 subjects. The atlas was constructed by normalizing all images to MNI-space, manually identifying the arteries and creating an average over the volume. The identification is done by extracting a vascular skeleton from the image, transforming it to MNI-space, labeling it with the atlas and transforming it back to subject space. The method was evaluated on a pilot cohort of 8 subjects. The rate of correctly identified arteries was 97%.**

Keywords— **Cerebral arteries, automatic labeling, 4D flow MRI, cerebrovascular atlas**

Introduction

The Circle of Willis (CoW) is an anastomititc structure of arteries at the skull base that connects the anterior and posterior cerebral circulation (figure 1). If one of the arteries are blocked or narrowed, it is possible for the blood to take an alternative path through this structure; therefore it is an interesting target for flow analysis.

Cerebral blood flow is usually measured with transcranial doppler or 2D PCMRI. These techniques can only measure flow in a predefined double-oblique plane, and defining these plains requires a lot of experience and knowledge of the vascular anatomy.

This study utilities a 4D flow MRI technique called PC VIPR [1] (phase contrast vastly undersampled isotropic projection reconstruction), which is a time resolved phase contrast technique where the entire brain is imaged within ten minutes, using radial sampling of k-space. This technique is not yet commercially available, and post-processing tools need to be developed.

There have been some different takes on making an automatic identification method for the cerebral arteries. One study models the arteries around the CoW as a rooted graph, divided into bifurcations. These bifurcations are labeled using a maximum a posteriori probability [2].

There has also been some work on cardiovascular atlases, for example, a fully labeled 3D interactive atlas of the human cerebrovascular system, containing 60 different variants, including their prevalence rate [3].

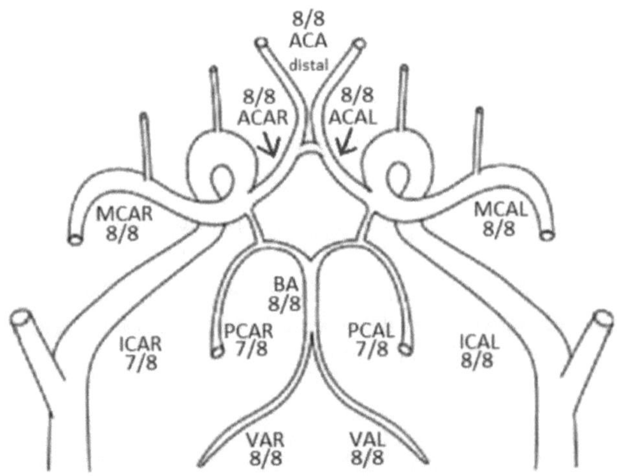

Fig. 1 Schematic image of the arteries included in the atlas. Left and right internal carotid artery (ICAL, ICAR), left and right vertebral artery (VAR, VAL), basilar artery (BA), left and right posterior cerebral artery (PCAL, PCAR), left and right middle cerebral artery (MCAR, MCAL), left and right anterior cerebral artery (ACAR, ACAL), distal anterior cerebral artery (ACAdistal). The numbers specifies the portion of the arteries that were correctly identified.

The atlas used in this study is developed in MNI-space, which means that the arteries are transformed to match a template based on 152 normal brains [4][5].

The aim of this study is to develop and evaluate an automatic atlas-based method for labeling of twelve arteries forming the CoW, for application in flow measurements.

Method

One of the key elements in this method was a diffeomorphic image registration algorithm called DARTEL [6] implemented in the spm8 toolbox (*http://www.fil.ion.ucl.ac.uk/spm*). This method was used to transform the images to MNI-space where the identification was done.

Another key element was the vascular tree construction method, which generates a single-voxel-wide centerline representation of the vascular tree, called the vascular skeleton [7].

By labeling the voxels in this skeleton and reconstructing the vascular tree from it, the arteries in the original image could be identified.

H. Mindedal and M. Persson (eds.), *16th Nordic-Baltic Conference on Biomedical Engineering*, IFMBE Proceedings 48, DOI: 10.1007/978-3-319-12967-9_11

SUBJECTS

For atlas construction, 25 subjects (15 men, 10 women, age 66.4±1.8 years) without any history of neurological diseases or stroke, and where all twelve arteries were visible and fully connected were included. For method evaluation, an additional 8 subjects (4 men, 4 women, age 65.3±0.7 years) with the same inclusion criteria was utilized.

MRI

The subjects were imaged with 4D flow MRI on a 3 Tesla General Electric HD750 system with a 32 channel headcoil. The field of view was 220 x 220 x 220 mm^3 with a 0.7 mm^3 isotropic resolution. The scan time was 9.5 min and the velocity encoding 110 cm/s.

ATLAS CONSTRUCTION

Twelve arteries were used for this atlas; left and right internal carotid artery (ICAL, ICAR), left and right vertebral artery (VAR, VAL), basilar artery (BA), left and right posterior cerebral artery (PCAL, PCAR), left and right middle cerebral artery (MCAR, MCAL), left and right anterior cerebral artery (ACAR, ACAL), and distal anterior cerebral artery (ACAdistal). The included arteries are shown in figure 1.

In addition to the arteries mentioned, the CoW consists of the anterior communicating artery, connecting the ACAR and ACAL, and the left and right posterior communicating arteries connecting each ICA to the corresponding PCA.

After data reconstruction, the images were normalized to a group template using DARTEL. The group template was created from probability maps for white matter, grey matter and CSF, constructed in spm8's *New Segment*. The probability maps were derived from the magnitude image provided by the 4D flow scan (presenting some T1 weighting). After normalizing to the group template, a second transformation was performed to normalize the images to MNI-space.

The subsequent processing was done using an inhouse tool developed in Matlab (*Mathworks*, MA, USA). A binary flow image was constructed by thresholding the velocity data. The threshold used to separate vessels from background was 18% of the maximum intensity value in the image [8].

The binary flow image was gradually thinned, using the vascular tree construction method [7] until a one voxel thick skeleton was obtained. This vascular skeleton was then pruned to remove loops and branches shorter than eight voxels. The next step was to label the voxels in the skeleton according to their location (branch point, end point or junction point), and label all branches for identification purpose.

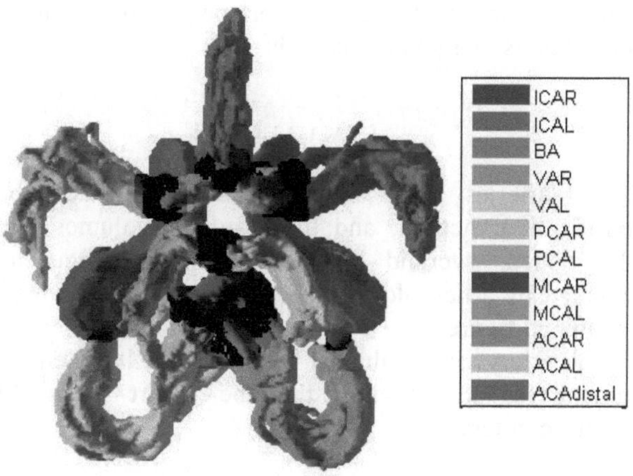

Fig. 2 Visualization of the 3D atlas used for arterial identification. The atlas is composed of 12 cerebral arteries from 25 subjects. The regions where the atlases of the arteries overlap are marked in black.

The arteries of interest were identified using a GUI where the binary flow image was visualized, and each artery was manually identified by clicking at them. If an artery by mistake was divided into more than one segment, all segments were joined together manually. This resulted in a binary map for each artery and subject. When needed, the MCA and PCA were shortened, to get a more cohesive atlas. The length of these segments depended both on flow rate and on branching pattern.

Each vascular tree was approved by a neurologist before the atlas was created.

The atlas consisted of a 3D volume containing spatial probabilities for each artery. These volumes were created by adding together all the binary volumes for each artery, and dividing by the number of subjects. This resulted in a volume with values between zero and one for each voxel in the volume. The resulting atlas is shown in figure 2.

IDENTIFICATION METHOD

The arterial identification was done by extracting the vascular skeleton the same way as for the atlas construction. The skeleton was then transformed to MNI space using the group template from the atlas construction. In MNI-space, each voxel in the vascular skeleton was assigned a value for each of the twelve arteries at the corresponding coordinate in the atlases.

The labeled skeleton was then transformed back to the native space using an inverse transform from the Data Processing Assistant for Resting-State fMRI (DPARSF) package from the R-fMRI network (*http://rfmri.org/)*, where the templates and flow-fields are used to transform the image from MNI-space back to the original coordinate system.

The shape of the arteries was recreated by dilating the identified segments of the vascular skeleton and masking them with the original binary image.

The last step before evaluation was to remove all arteries shorter than eight voxels (5 mm). In the areas where the identified arteries overlaped, it was adjusted with a nearest neighbor approach.

EVALUATION

The arteries were labeled using the described identification method, and the resulting volumes were color coded, overlaid on the magnitude image, and evaluated by a neurologist. An example of these images is shown in figure 3.

Each artery was evaluated based on if a large enough segment was labeled, and if the labeled part belonged to the correct artery.

EVALUATION CRITERIA

To say that a large enough segment was labeled, some basic criteria were set up.

ICA: a 2 cm segment of the cervical segment

VA: a 1 cm straight segment proximal or distal to the foramen magnum

BA: a 1 cm segment anywhere in the artery

PCA: a 1 cm segment distal to the posterior communicating artery (P2)

MCA: a 1 cm segment in the proximal part of the artery (M1) as well as in a 1 cm segment in the sylvian fissure.

ACA: a 5 mm segment, proximal to the anterior communicating artery (A1).

ACAdistal: a 1 cm segment in the distal part of the arteries.

In addition to this, the presence of a P1 segment (PCA proximal to the posterior communicating artery) and the proximal part of the ICA was noted in the evaluation, without any length requirements.

Results

93 of 96 arteries (97%) were correctly identified. In one subject, the distal part of the ICAR, as well as the left P2 segment was not detected. In another subject, the method failed to identify the right P1 segment. Two ACAR and two BA were borderline short, but still fulfilled the predetermined criteria.

In two cases, the artery identified as the VA included parts of an artery with close proximity, but the main part of the identified volumes were still correct. These results are summarized in figure 1.

Discussion

These results are based on a small pilot group, with all twelve large cerebral arteries present, but show promise for the future.

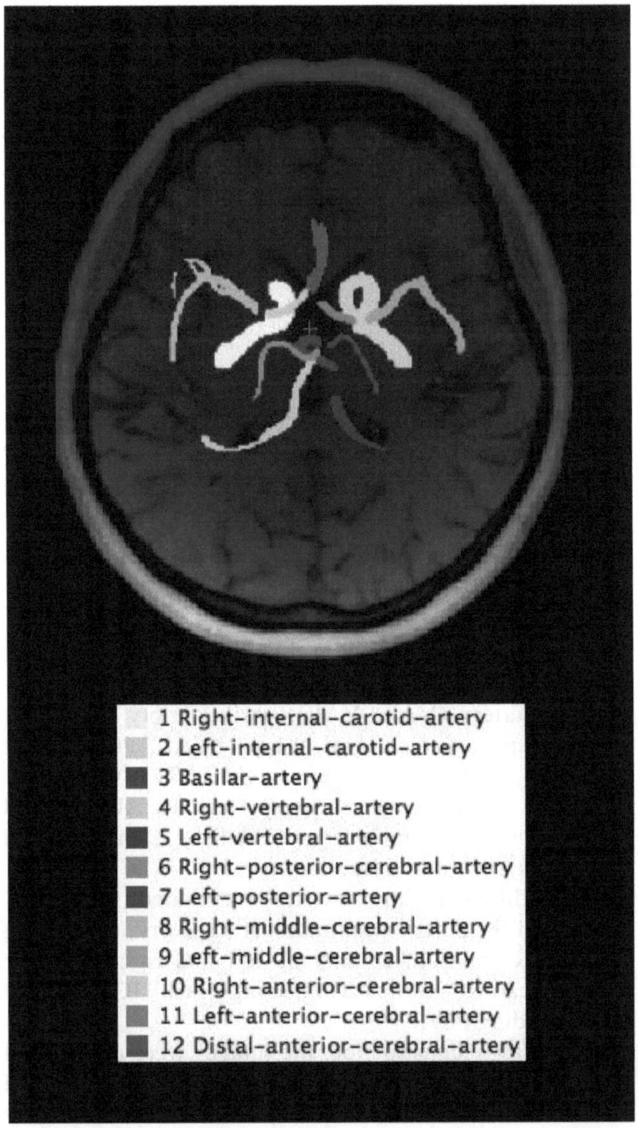

Fig. 3 Example of a labeled vascular tree overlaid on the magnitude image. In this subject all arteries were approved.

EVALUATION CRITERIA

The segments defined in the evaluation criteria are meant to be used for flow measurements. The different lengths of the segments depend on the length of the arteries. A longer segment gives a more stable measurement, 1 cm is a suitable length for most vessels, but ACA is too short to get a segment of that length even with a perfect identification. ICA on the other hand, is long enough to demand a longer segment for evaluation.

This doesn't really agree with the fact that smaller arteries require longer segments to get the same signal to noise ratio as larger arteries.

The segments that don't have any length requirements are not planned to be subject for flow measurements, but still important to get a correct identification.

METHOD CONSIDERATIONS

One problem with this method is that, due to the empirical thresholding, arteries with low velocities are hard to detect. This can be seen by comparing the anatomical images to the binary flow images.

The threshold was set to the default value. A higher threshold value would result in thinner and fewer arteries, while a lower threshold gives more and thicker arteries. This would results in a vascular skeleton with more loops and branches.

Low velocity in relation to the velocity sensitivity is probably the cause of the two visually incomplete PCAs too. In the subject with the unlabeled P1 segment, a high flow in the corresponding posterior communicating artery was seen, which indicates that a larger part of the blood in the P2 segment origins from that pathway, resulting in a lower blood flow in the P1 segment.

The incomplete labeling of the ICAR was most likely caused by a gap in the labeling of the vascular skeleton. When recreating the vessel from the labeled skeleton, the artery is identified as the longest continuous segment with the corresponding label. If one point in the branch has another label, the branch will be divided into two parts and only the longest part will be identified as the artery in question. This is done to avoid incorrect classification of proximal arteries, and therefore improve the accuracy.

As exemplified in figure 3, the part of the vascular tree that was hardest to identify was the part where the two VA join together to form the BA. This area has large variation between individuals and the region of overlap between these arteries in the atlas is quite large, as seen in figure 2.

ANATOMICAL VARIATIONS

There are a variety of possible anatomies for the CoW. A common malformation of is the fetal PCA, which means that the P1 segment is missing, and all the blood flow through the PCA origins from the ICA. It is also not uncommon to have a missing or underdeveloped A1 segment [9].

The next step in this work is to evaluate the method in a larger cohort with more deviant morphologies, and to add the posterior communicants to the atlas, primarily to handle the fetal types. The anterior communicating artery is, in the cases it could be separated from the two ACA, included in the ACAdistal atlas.

CONCLUSION

A cerebrovascular atlas was developed from 4D flow MRI on 25 subjects. This atlas was used to identify and label twelve arteries forming the CoW.

The method was evaluated on eight subjects. 93 of 96 arteries were correctly identified, and only two of the identified arteries included any incorrect segment. This means that both the specificity and the sensitivity for the method were high.

There are some minor problems with the method, but the results are very promising. The next step in this work is to further develop the atlas by including more arteries, and to evaluate the method for a wider range of subjects.

ACKNOWLEDGMENT

Swedish Research Council (grant number 221-2011-5216), The Swedish Heart-Lung Foundation.

REFERENCES

1. Gu T, Korosec F R, Block W F et al. (2005) PC VIPR: A High-Speed 3D Phase-Contrast Method for Flow Quantification and High-Resolution Angiography. Am J Neuroradiol 26:743–749
2. Bogunović H, Pozo J M, Cárdenes R et al. (2013) Anatomical Labeling of the Circle of Willis using Maximum A Posteriori Probability Estimation. IEE Trans Med Imaging. 32:1587-1599 DOI 10.1109/TMI.2013.2259595
3. Nowinski W L (2009) A Three-Dimensional Interactive Atlas of Cerebral Arterial Variants. Neuroinform 7:255-264 DOI 10.1007/s12021-009-9055-0
4. Cutini S, Scatturin P, Zorzi M (2001) A new method based on ICBM152 head surface for probe placement in multichannel fNIRS. NeuroImage 54:919-927 DOI 10.1016/j.neuroimage.2010.09.030
5. Evans A C, Mills S R, Brown E D et al. (1993) 3D statistical neuroanatomical model from 305 MRI volumes. IEEEE Proc. Vol. 3 Nucl Sci Symp Med Imaging Conf, San Francisco, CA, 1993 DOI 10.1109/NSSMIC.1993.373602
6. Ashburner J (2007) A fast diffeomorphic image registration algorithm. NeuromImage 38:95-113 DOI 10.1016/j.neuroimage.2007.07.007
7. Chen Z, Molloi S (2003) Automatic 3D vascular tree construction in CT angiography. Comput Med Imaging Graph 27:469-479 DOI 10.1016/S0895-6111(03)00039-9
8. Wåhlin A, Ambarki K, Birgander R et al. (2013) Measuring Pulsatile Flow in Cerebral Arteries Using 4D Phase-Contrast MR Imaging. Am J Neuroradiol 34:1740-1745 DOI 10.3174/ajnr.A3442
9. Krabbe-Hartkamp, M J, van der Grond, J, de Leeuw, F E et al. (1998) Circle of Willis: morphologic variation on three-dimensional time-of-flight MR angiograms. Radiology 207:103-111

Author: Tora Dunås
Institute: Umeå University
Street:
City: Umeå
Country: Sweden
Email: tora.dunas@radfys.umu.se

HRV Analysis in CHF Patients with Different Etiologies by Means of Spectral and Nonlinear Indexes

E. Fornasa[1], A. Accardo[1], M. Cinquetti[2], M. Merlo[2], and G. Sinagra[2]

[1] Dept. of Engineering and Architecture, University of Trieste, Trieste, Italy
[2] Cardiovascular Dept., Ospedali Riuniti and University of Trieste, Trieste, Italy

Abstract— **Heart rate variability (HRV) has been widely investigated in a variety of clinical conditions, but few studies have assessed the changes in the heart rate with regard to different congestive heart failure (CHF) etiologies. The aim of this work was to characterize the heart rate variability of patients affected by CHF caused by dilated cardiomyopathy, ischemic heart disease or other different CHF etiologies, in comparison to that of healthy subjects. The parameters adopted for analysis were the power spectral density in the very low, low and high frequency bands, the power-law β exponent and the SD1 and SD2 indexes of the Poincaré plot. Results showed significant changes related to etiology especially in the very low and high frequency bands and in the β exponent values, thus encouraging the future development of a classifier for the discrimination of the different CHF etiologies.**

Keywords— **Heart rate variability, congestive heart failure, etiology, spectral analysis, nonlinear indexes.**

I. INTRODUCTION

Heart rate variability (HRV) is a widely investigated signal that provides diagnostic and prognostic information in a variety of clinical conditions [1]. In particular, HRV has been extensively studied in cardiac patients to assess the sympathovagal balance within the cardiovascular system. Reduced or abnormal HRV patterns have been associated to the severity of the congestive heart failure (CHF) and to an increased risk of mortality [2], but few studies have assessed the changes in the heart rate in reference to different CHF etiologies [3, 4].

The purpose of this work was to characterize the HRV of patients affected by CHF caused by (1) ischemic heart disease, (2) dilated cardiomyopathy (DCM) or (3) other different CHF etiologies (e.g. hypertensive and valvular heart disease). At this end, some spectral and nonlinear HRV indexes were calculated for the three groups of patients with different CHF etiologies and compared with those of a control group composed by healthy subjects.

Several parameters have been proposed to analyze the variations in the RR interval time series in both time and frequency domain, and also numerous nonlinear methods have been suggested to give a better insight of the nonlinear contributions of the autonomic control [1]. The parameters used in this work were the power spectral analysis in very low, low and high frequency bands, the power-law β exponent and the SD1 and SD2 indexes of the Poincaré plot.

II. MATERIAL AND METHODS

A. Study population and electrocardiographic monitoring

The study population was composed of 75 subjects divided into three equal groups, matched for age and gender, depending on CHF etiology. The first group (age 66±10, 80% males) was composed by patients with ischemic heart disease, the second (age 63±10, 80% males) by patients with dilated cardiomyopathy, and the third (age 63±5, 80% males) by patients with other CHF etiologies. A forth group, composed by 25 normal healthy subjects (age 63±9, 80% males) with no history of heart disease, was considered as control.

All the subjects underwent a 24-hours Holter monitoring using a 3 channels time-tracking system. RR intervals were automatically extracted from ECG records using SyneScope® Holter analysis software (Sorin Group, Italy), which also labelled each RR interval as normal, ectopic or artifact.

B. RR time series pre-processing

Being HRV influenced by circadian variations [5], two time epochs lasting five hours each in day and night-time were considered for each subject. Day-time was defined from 3 pm to 8 pm and night-time from 12 pm to 5 am.

Each RR time series included a percentage of artifacts lower than 5% and the duration of the longest artifact segment was lower than 10 seconds. This allowed obtaining a reliable estimation of the HRV parameters because correction, which is a possible cause of significant errors in the estimation of HRV indices, was applied on a minimal part of the RR intervals [6, 7]. In this work the ectopic beats, commonly considered as a source of error in the HRV analysis [8], were included in the evaluation because of their relevant role in the discrimination of the CHF etiology [8-10].

H. Mindedal and M. Persson (eds.), *16th Nordic-Baltic Conference on Biomedical Engineering*,
IFMBE Proceedings 48, DOI: 10.1007/978-3-319-12967-9_12

Cubic spline interpolation based on the normal and ectopic RR intervals was applied and the non-evenly spaced in time RR sequences were subsequently resampled at 2 Hz to obtain a constant sampling time.

C. HRV analysis: spectral and nonlinear indexes

To analyze HRV in the frequency domain, the power spectral density (PSD) of the edited RR time series was calculated using Welch's periodogram method. The time series were segmented into sections of 512 seconds each, windowed with a Hamming window, with 50% overlap. Three typical spectral bands were distinguished [11]: Very Low Frequency (VLF) < 0.04 Hz, Low Frequency (LF) 0.041–0.15 Hz and High Frequency (HF) 0.151–0.4 Hz. Spectral components were expressed in normalized units (n.u.), which represent the relative value of each component in relation to the total power.

For the analysis of the fractal properties of the heart rate, the power versus frequency relationship was investigated in a log-log plot [12]. The power-law β exponent was calculated as the slope of the regression line fitting the PSD with the sign inverted (Figure 1).

Furthermore, the RR time series were investigated from a geometrical and nonlinear point of view using the Poincaré plot technique. For the quantitative analysis of the plot, an ellipse was fitted to the points of the chart and the SD1, i.e. the minor axis of the ellipse, and the SD2, i.e. the major axis of the ellipse, were calculated. The SD1 is an index of the instantaneous variability of the RR time series, whereas the SD2 represents the long-term HRV and reflects the overall variability [13].

D. Statistical analysis

Since normal Gaussian distribution of the data was rejected by the Lilliefors test at the 5% significance level, the nonparametric Kruskal-Wallis test was used to compare the four groups (normal, ischemic, DCM and other CHF etiology). Furthermore, Wilcoxon rank sum test was used to compare each pair of groups, followed by Bonferroni's correction due to multiple testing. Differences were considered significant for a p-value < 0.05. Medians with 25th and 75th percentile were calculated for each parameter and each group.

III. Results

Table 1 presents the median values of the proposed parameters calculated for each group. With regard to the VLF,

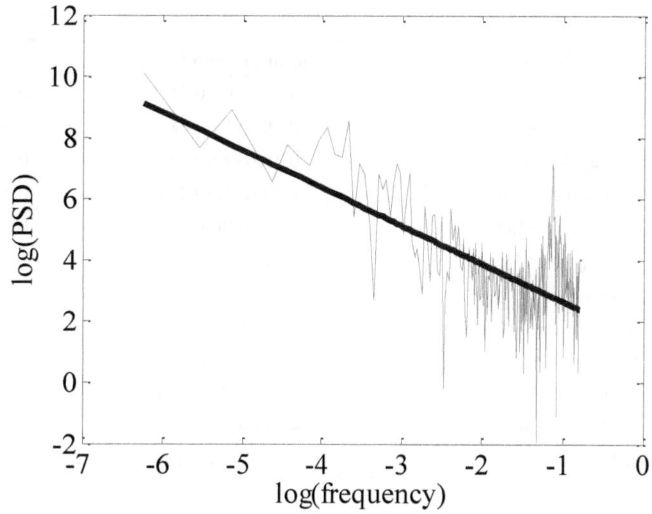

Fig. 1 Example of a log-log plot of the periodogram calculated from a RR time series of a normal subject. The power-law β exponent is the slope of the regression line fitting the data. In this case, $\beta = 0.8$.

both in day-time and night-time, the DCM group shows intermediate values between the normal group and the ischemic one, with the last one having significant lower median values with respect to all the other groups (Table 2). The group of CHF patients with etiology different from both DCM and ischemic presents VLF values intermediate between the normal and the DCM groups, without statistically significant difference. Only in the night-time, the VLF is significantly different between the normal and the DCM groups. Similarly to what happens for the VLF values, also the HF values of the DCM group are intermediate to the ones of the normal and ischemic groups, with the ischemic group having higher median values both in day-time and night-time. The differences are significant among the normal, DCM and ischemic groups, while the group with patients with different CHF etiologies has once again values intermediate and partially overlapping those of the normal and the DCM groups.

In the day-time, the same significant differences of the HF values are observed also for the β values, with higher medians for the normal and progressively lower for the DCM and the ischemic groups. In the night-time, the β exponent shows significant differences between all the pairs of groups except between the group with various CHF etiologies and the DCM one.

The parameter SD1 presents a high variability among the subjects of each group, both in day-time and night-time, so determining a reduction in the number of significant differences of the median values between the groups. Nevertheless, in the day-time the ischemic group has significant different values from both the normal and the group with

Table 1 Median (25th-75th percentile) of the different parameters for each group.

Parameter	Normal (control)	DCM	Ischemic	Other CHF etiology
VLF day-time [n.u.]	0.68 (0.61-0.75)	0.55 (0.45-0.69)	0.29 (0.18-0.45)	0.62 (0.53-0.76)
VLF night-time [n.u.]	0.69 (0.64-0.78)	0.55 (0.38-0.67)	0.26 (0.14-0.43)	0.64 (0.48-0.73)
LF day-time [n.u.]	0.18 (0.12-0.23)	0.15 (0.14-0.19)	0.18 (0.13-0.24)	0.17 (0.13-0.21)
LF night-time [n.u.]	0.19 (0.12-0.22)	0.18 (0.16-0.21)	0.17 (0.11-0.24)	0.15 (0.14-0.19)
HF day-time [n.u.]	0.12 (0.06-0.18)	0.26 (0.13-0.37)	0.56 (0.35-0.63)	0.18 (0.10-0.28)
HF night-time [n.u.]	0.12 (0.07-0.16)	0.24 (0.13-0.43)	0.55 (0.35-0.69)	0.18 (0.10-0.31)
β day-time	1.27 (0.93-1.54)	0.83 (0.47-1.11)	0.24 (0.03-0.62)	0.97 (0.58-1.32)
β night-time	1.41 (1.10-1.53)	0.85 (0.39-1.23)	0.26 (-0.03-0.68)	1.00 (0.70-1.32)
SD1 day-time	21.41 (14.98-27.85)	25.68 (20.35-32.26)	40.47 (26.79-55.80)	23.60 (17.49-33.13)
SD1 night-time	21.71 (17.02-31.46)	31.37 (23.89-59.68)	50.68 (28.94-67.46)	23.12 (16.27-45.52)
SD2 day-time	68.13 (49.43-81.59)	62.31 (49.29-83.45)	78.50 (51.96-108.45)	65.09 (50.87-84.63)
SD2 night-time	68.13 (57.86-106.98)	89.71 (60.17-132.59)	95.59 (67.19-121.15)	73.43 (51.51-114.21)

different CHF etiologies, whereas in the night-time the control group has significantly different values from both the ischemic group and the DCM one. Finally, no significant difference among the groups is present for the LF and SD2 parameters, both in day and night time.

IV. Discussion

This study aimed at evaluating if the proposed spectral and nonlinear parameters could effectively separate the different CHF etiologies. For this purpose, given that the time series including premature beats seems to provide more powerful prognostic information [8], the replacement of ectopic beats, which is traditionally applied to evaluate the autonomic nervous system activity, was avoided. The inclusion in the HRV analysis of ectopic beats causes higher instantaneous irregularities of the RR series and an increase in the randomness of the RR dynamics, thus determining lower HF values in normal subjects with respect to the DCM and the ischemic groups. In fact, the occurrence of ectopic beats is usually higher in CHF patients than in healthy subjects.

At the same time, the inclusion of ectopic beats makes the present results not comparable to other studies in which ectopic beats were excluded from the RR time series (e.g. [14], which presents higher HF values for normal subjects with respect to CHF ones). Moreover, the physiological interpretation of the spectral components cannot be easily associated to the sympathetic and parasympathetic activity of the autonomous nervous system. In fact, it is known that ectopic beats introduce a bias into HRV measures representing an obstacle in the interpretation of the results [6, 8].

Table 2 P-values of the nonparametric Wilcoxon rank sum test followed by Bonferroni's correction for each pair of groups.

Parameter	Normal vs DCM	Normal vs Ischemic	Normal vs other CHF etiology	DCM vs Ischemic	DCM vs other CHF etiology	Ischemic vs other CHF etiology
VLF day-time	n.s.	<2E-06	n.s.	<0.001	n.s.	<6E-05
VLF night-time	<0.0006	<4E-08	n.s.	<0.002	n.s.	<4E-06
LF day-time	n.s.	n.s.	n.s.	n.s.	n.s.	n.s.
LFnight-time	n.s.	n.s.	n.s.	n.s.	n.s.	n.s.
HF day-time	<0.03	<9E-07	n.s.	<0.002	n.s.	<4E-05
HF night-time	<0.005	<3E-08	n.s.	<0.002	n.s.	<8E-06
β day-time	<0.02	<9E-07	n.s.	<0.003	n.s.	<0.0003
β night-time	<0.0005	<9E-08	<0.01	<0.004	n.s.	<9E-05
SD1 day-time	n.s.	<0.002	n.s.	n.s.	n.s.	<0.04
SD1 night-time	<0.02	<0.0005	n.s.	n.s.	n.s.	n.s.
SD2 day-time	n.s.	n.s.	n.s.	n.s.	n.s.	n.s.
SD2 night-time	n.s.	n.s.	n.s.	n.s.	n.s.	n.s.

The intermediate values found in the group of patients with different CHF etiologies in respect of the other groups, can be explained by the high heterogeneity of the clinical conditions of this group. More interesting are the intermediate values of many parameters found in the DCM group with respect to both normal and ischemic groups. This finding requires further analysis refining of this group in order to identify the clinical characteristics which determine similarities as well as dissimilarities from the normal and the ischemic groups.

The VLF, the HF and the β exponent have shown good results with significant differences between most of the pairs of groups, both in day-time and night-time. In particular, the β exponent in the night-time performed better than the other parameters and therefore it can be considered a good indicator of the CHF etiology.

V. Conclusions

This work demonstrated that spectral and nonlinear HRV measures could be successfully used in the HRV feature identification of patients with different CHF etiologies and healthy subjects.

The group with DCM patients and the one with different etiologies require deeper investigations to punctually define the clinical condition of the subjects, thus obtaining more homogeneous groups. Nevertheless, considering the statistical significance of some of the proposed HRV parameters, further studies will concern the development of a classifier for the discrimination of the different CHF etiologies starting from the spectral and nonlinear indexes presented in this work.

Acknowledgment

Work partially supported by University of Trieste, Master in Clinical Engineering.

References

1. Rajendra Acharya U, Paul Joseph K, Kannathal N et al (2006) Heart rate variability: a review. Med Biol Eng Comput. 44:1031-1051.
2. Huikuri HV, Stein PK (2013) Heart rate variability in risk stratification of cardiac patients. Prog Cardiovasc Dis 56:153–159 DOI 10.1016/j.pcad.2013.07.003
3. Voss A, Schroeder R, Vallverdú M. et al (2013) Short-term vs. long-term heart rate variability in ischemic cardiomyopathy risk stratification. Front Physiol. 4:364 DOI 10.3389/fphys.2013.00364
4. Ponikowski P, Anker SD, Chua TP et al (1997) Depressed heart rate variability as an independent predictor of death in chronic congestive heart failure secondary to ischemic or idiopathic dilated cardiomyopathy. Am J Cardiol. 79:1645-50.
5. Accardo A, Cusenza M, De Felice A et al. (2012) Ultradian rhythms during day and night in normal and COPD subjects. Stud Health Technol Inform. 180:1120-2.
6. Haaksma J, Dijk WA, Brouwer J et al. (1995) Effects of automatic ectopy exclusion on the analysis of Heart Rate Variability using a percentile exclusion rule. Proc. of the 1995 Conference on Computers in Cardiology, Vienna, Austria, 197-200
7. Mateo J, Laguna P (2000) Improved heart rate variability signal analysis from the beat occurrence times according to the IPFM model. IEEE Trans Biomed Eng 47:985-996.
8. Peltola MA, Seppanen T, Makikallio TH et al (2004) Effects and Significance of Premature Beats on Fractal Correlation Properties of R-R Interval Dynamics. Ann Noninvasive Electrocardiol 9:127-135
9. Huikuri HV, Makikallio TH, Peng CK et al (2000) Fractal correlation properties of R-R interval dynamics and mortality in patients with depressed left ventricular function after an acute myocardial infarction. Circulation 101:47-53.
10. Vikman S, Makikallio TH, Yli-Mäyry S et al (1999) Altered complexity and correlation properties of R-R interval dynamics before the spontaneous onset of paroxysmal atrial fibrillation. Circulation 100:2079-84.
11. Task Force of the European Society of Cardiology and the North American Society of Pacing and Electrophysiology (1996) Heart rate variability: standards of measurement, physiological interpretation and clinical use. Circulation 93:1043–65
12. D'Addio G, Corbi G, Accardo A et al (2009) Fractal behaviour of heart rate variability reflects severity in stroke patients. Stud Health Technol Inform. 150:794-798.
13. Hoshi RA, Pastre CM, Vanderlei LC et al (2013) Poincaré plot indexes of heart rate variability: relationships with other nonlinear variables. Auton Neurosci. 177:271-274. DOI 10.1016/j.autneu.2013.05.004.
14. Wu GQ, Arzeno NM, Shen LL et al. (2009) Chaotic signatures of heart rate variability and its power spectrum in health, aging and heart failure. PLoS One 4:e4323. DOI 10.1371/journal.pone.0004323

Corresponding author:

Author: Elisa Fornasa
Institute: University of Trieste, Dept. of Engineering and Architecture
Street: Via A. Valerio 10
City: Trieste
Country: Italy
Email: elisa.fornasa@phd.units.it

A Novel Model for Screening Aortic Stenosis Using Phonocardiogram

Arash Gharehbaghi[1,2], Per Ask[1], Maria Lindén[2], and Ankica Babic[1,3]

[1] Physiological Measurements, Department of Biomedical Engineering, Linköping University, Linköping, Sweden
[2] Division of Intelligent Future Technologies, Biomedical Engineering Group, Mälardalen University, Västerås, Sweden
[3] Department of Information Science and Media Studies, University of Bergen, Norway

Abstract— **This study presents an algorithm for screening aortic stenosis, based on heart sound signal processing. It benefits from an artificial intelligent-based (AI-based) model using a multi-layer perceptron neural network. The AI-based model learns disease related murmurs using non-stationary features of the murmurs. Performance of the model is statistically evaluated using two different databases, one of children and the other of elderly volunteers with normal heart condition and aortic stenosis. Results showed a 95% confidence interval of the high accuracy/sensitivity (84.1%-86.0%) /(86.0%- 88.4%), thus exhibiting a superior performance to a cardiologist who relies on the conventional auscultation. The study suggests including the heart sound signal in the clinical decision making due to its potential to improve the screening accuracy.**

Keywords— **Aortic stenosis, phonocardiography, heart sound, heart murmurs.**

I. INTRODUCTION

Rapid technological progresses have enabled new medical devices for enhancing the decision making using advanced signal processing techniques. The sophistication of signal analysis equipment has allowed extracting information even from the noisy signals seemingly difficult to process. However, screening of the heart diseases is often performed according to the subjective decision of the nurses or practitioners in the primary healthcare centers, where the patients assessed as having pathological changes are sent to the hospitals to undergo echocardiography. Studies have showed that the screening accuracy in the primary healthcare centers is still low [1], causing a large number of the non-pathological patients to be sent to specialized wards, and to a lesser extent some pathological cases to be ignored. As a consequence, the cost benefit of the clinical assessments may be unreasonably declined. Heart sound auscultation has been used as a screening technique for cardiac diseases over centuries, but the main breakthrough came up in 1816 when the first stethoscope was created [2]. Phonocardiogram is a plot of the heart sounds on the rolled paper that was used as a part of the patient's record with the diagnostic value [3]. However, it turned out to become an obsolete technique after development of the echocardiography. One reason of this discontinuation was its poor signal quality reflected on the plots as the extra and ambiguous points. In the recent decades, rapid progresses in electronic devices offered high quality electronic stethoscopes by which the heart sound signal could be collected and electronically incorporated into the patient records to facilitate cardiac assessments. A heart sound recording is normally composed of two essential sounds in each cycle, named the first and second heart sound, S1 and S2, respectively [2]. Murmurs are extra sounds that could be in systolic or diastolic phases, acting as the main group of the extra sounds. Clicks, friction rubs, opening snap, venous hum, third heart sound and forth heart sound are the other extra sounds. Valvular heart diseases are often manifested by the extra heart sound associated with them. Aortic Stenosis (AS), tightening of the aortic valve orifice, is common valvular heart disease whose prevalence is increased in elderly people [4]. This disease could have a long asymptomatic period (up to 20 years) with no physical symptom, after which the average life time is estimated to be 5 years where sudden death is seen in 68% of the patients during the first year [5][6][7]. A diamond shape systolic murmur whose intensity is consistent with respiration, is a direct symptom of the AS, which could be appeared even in the early stages of the defect when its physical manifestation is still hidden [2][4].

In this study, we present an algorithm for screening aortic stenosis to enable an accurate diagnostics of heart sound signal. Signal analysis is done using our novel model which employs Artificial Intelligence (AI) techniques. It is assessed in the case study with elderly patient material due to the screening complexity and high disease prevalence in elderly people. With minor modifications, the AI based model could be easily adopted for screening of other types of the heart murmurs.

II. METHODS

A. *The proposed system*

The algorithm has an ability to record and display the heart sound signal to support decision making. It screens and analyzes the systolic murmur over a set of the cardiac cycles to detect the presence of aortic stenosis. Heart sound signal is recorded at the chest of the patient, using an electronic stethoscope with the link to a computer which is commercially available on the market. The recorded signal is digitally processed by the AI algorithm to detect and classify the murmurs. The outcome is displayed at the convenience of a physician. A portable computer or even a smart phone can serve as the processing platform for the screening purposes where the recordings together with the other patient demographic and clinical data, are electronically stored as a part of an electronic patient record. Such a processing platform allows users to exploit the patient records even in more demanding clinical settings

H. Mindedal and M. Persson (eds.), *16th Nordic-Baltic Conference on Biomedical Engineering*,
IFMBE Proceedings 48, DOI: 10.1007/978-3-319-12967-9_13

such as tele-medicinal.

B. The AI-based model

The proposed intelligent analysis is based on modeling time-frequency contents of the systolic murmur caused by aortic stenosis. The recorded heart sound signals are firstly filtered and segmented using conventional signal processing techniques by which beginning and end of the cardiac cycles, as well as the S1 and S2 segments, are identified. Details of the automatic algorithms for heart sound segmentation could be found in [8] and [9]. Our model employs the systolic phase between the end of the S1 and beginning of the S2. The model is based on the joint combination of the artificial neural network and the statistical analysis, conjunctly preserving dynamic characteristics of the signal both within the systolic phase and over the cycles. The Figure 1 illustrates block diagram of the model.

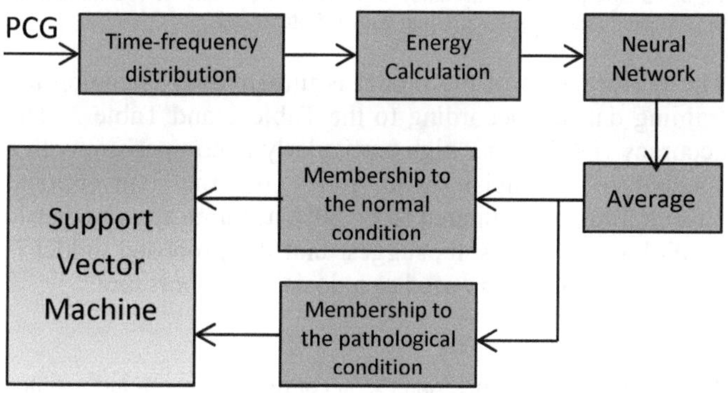

Figure 1 The processing model for diagnosing diseases related changes i.e. AS.

The systolic murmur is characterized by its frequency contents over a set of the temporal windows with fixed length, spanning whole the systolic duration in time. Spectral powers of each window over a number of frequency bands, named discriminating frequency bands, are calculated to be employed by the model as the inputs. An artificial neural network is trained for learning murmur characteristics, using the spectral powers of the subsequent windows as the inputs. A multi-layer perceptron neural network, trained by the back propagation delay method, is empirically favored for this study due to its flexibility in learning. The number of the neurons in the hidden and output layers is 30 and 1, respectively. The activation function of the hidden layer is tangential sigmoid while for the output layer a logarithmic sigmoid is selected since its output is strictly to the real numbers between 0 and 1.

The output of the neural network is obtained for each individual cycle, and averaged over all the cycles of the recording. This procedure is applied on all the recordings of the training dataset, resulting in real numbers between 0 and 1 as the indicative numbers for the aortic stenosis (AS) and healthy classes. Then, statistical distributions of the indicative numbers are modeled by a Gaussian curve, separately for the healthy and AS related conditions. The membership of the indicative number to the AS or healthy class is estimated according to its likelihood of belonging to those two classes using the two statistical distributions. Then the 2D vectors constituted of the membership values, are employed for discriminating the individuals with the disease from the healthy ones.

C. Study population

Two databases consisted of heart sound recordings of 10-seconds duration, collected from the patients referred to the Linköping University Hospital and Tehran Children Medical Center, were subjected to the study. An electronic stethoscope of Welch Allyn Meditron was used along with a personal computer equipped with a 16-bit sound card and sampling rate of 44.1 KHz. The recordings were down-sampled to 2 kHz, using an anti-aliasing filter, for processing relying on the fact that the frequency contents of heart murmurs are mainly below 600 Hz which could be extended up to 1000 Hz in rare conditions [10].

All the referred patients underwent echocardiography and complementary cardiac tests according to the guidelines of the hospitals, which are in compliance with the World Medical Association. The study was approved by the appropriate ethic committees and complied with the Good Clinical Practices (GCP). Two heart conditions were considered in the inclusion criteria of the study; the aortic stenosis and the healthy conditions. Table 1 lists the basic characteristics of the patient population.

Table 1 Specification of the data materials

	Number of individuals with AS	Number of the healthy individuals	Age range (years)	Data collection place
Database 1	14	0	45-93	Linköping
Database 2	16	30	1-16	Tehran

D. The statistical evaluations

The system is statistically evaluated using repeated random sub-sampling with the balanced 50%/50% of training/testing rate. In this method, 50% of each database is randomly selected for training and the rest for testing the system. This procedure is repeated several times with different sets of training and testing data to verify significance of the results. The testing includes calculation of the three performance measures: accuracy (Γ_{AC}), sensitivity (Γ_{SE}) and specificity (Γ_{SP}), which are defined as follows:

$$\Gamma_{AC} = \frac{100\,(N_{TP}+N_{TN})}{N_{TP}+N_{TN}+N_{FP}+N_{FN}} \quad (1)$$

$$\Gamma_{SE} = \frac{100 N_{TP}}{N_{TP}+N_{FN}} \quad (2)$$

$$\Gamma_{SP} = \frac{100 N_{TN}}{N_{FP}+N_{TN}} \quad (3)$$

where N_{TP} and N_{TN} are the number of correctly classified signals for aortic stenosis and healthy classes, respectively. N_{FP} is the number of the healthy patients classified as aortic stenosis and N_{FN} is the number of the aortic stenosis classified as the healthy patients.

III. RESULTS

The membership values of the patients with aortic stenosis and healthy classes, described in the Section 2.2, are plotted for all recordings of the databases (Figure 2).

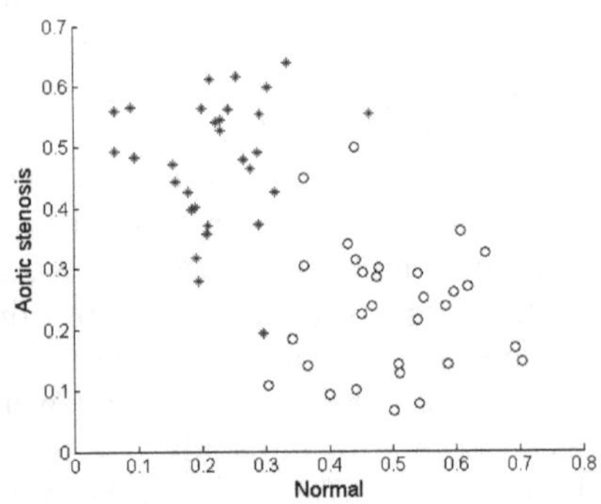

Figure 2 Membership values for the individuals with healthy (O) and aortic stenosis (*) conditions.

As seen in the figure above, the two classes are well discriminated by the AI-based model. In order to quantify the discrimination, the two classes are separated by using a support vector machine classifier. Details of the support vector machine technique could be found in [11] and [12]. Table 2 lists the resulting accuracy, sensitivity and specificity. The Figure 1 and Table 2 signify a high potential of the proposed model for an accurate AS screening. In order to investigate robustness of the model against the test data, repeated random sub-sampling procedure, was employed.

Table 2 The calculated accuracy, sensitivity and specificity of the AI-base model

	Accuracy (%)	Sensitivity (%)	Specificity (%)
Calculated value	95	97	93

The Figure 3 illustrates the box-plot of the accuracy, sensitivity and specificity for 100 iterations. The Table 3 represents their descriptive statistics in more detail. The confidence intervals are obtained using t-student test with the confidence level of 95%.

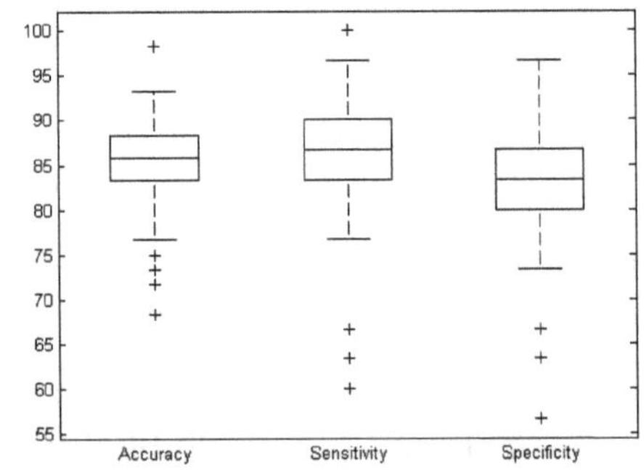

Figure 3 Box-plot of the accuracy, sensitivity and specificity, resulted from the repeated random sub-sampling with 100 iterations.

The performance of the model is impaired by changing the training dataset, according to the Table 2 and Table 3. The accuracy is still rather high particularly in comparison with a specialized physician who relies on the conventional auscultation, as estimated to be 80% accurate for a pediatric cardiologist [1]. Results suggest that the proposed model is accurate enough to be used as a decision support tool.

Table 3 Confidence intervals (95% confidence level) and resulted from the repeated random sub-sampling with 100 iterations.

	Confidence Interval
Accuracy	84.1 – 86.0
Sensitivity	86.0 – 88.4
Specificity	81.8 – 84.1

IV. DISCUSSIONS

This study suggests utilizing the heart sound signal as a part of the electronic patient record due to its potential to enhance the heart diseases screening. In addition, the here presented study proposes an AI-based model for a robust screening of the aortic stenosis over a broad range of the patients in terms of the age, which makes it unique as compared to other pertinent researches [13][14][15][16]. Previously, a number of the studies showed that a computer-assisted phonocardiogram can increase the screening accuracy in children [1][17][18][19]. On the other hand, serious studies employed the heart sound recordings as a diagnostic signal for different heart abnormalities, showing the importance and applicability of the heart sound recording [20][21][22][23]. In comparison, the results of above-represented study suggest the superiority of the model for screening the mild pathologies, raising its applicability in the clinical routines. The statistical analysis of the results reflects robustness of the model against the input data, which is accounted as another consolidating feature of the study.

The model could be trained to learn characteristics of the other pathological heart sounds for a comprehensive cardiac examination. Mitral insufficiency or congenital heart diseases are other abnormalities that could be screened by the proposed model.

The presented model is simply incorporable into a smart phone to be used as a home health care system with the telemedicine facility.

V. Conclusion

This study introduced a capable algorithm for screening aortic stenosis, the common cardiac disease in elderly people. The system employs a novel AI-based model to improve decision accuracy and sensitivity of the aortic stenosis screening. The statistical analysis showed high screening accuracy of (84.1%-86.0%) with 95% confidence level, exhibiting noticeable enhancement in comparison with the conventional auscultation.

References

1. R. L. Watrous, W. R. Thompson, S. J. Ackerman. The impact of computer-assisted auscultation on physician referrals of asymptomatic patient with heart murmurs. Clin. Cardiol. 2008;31:79–83.

2. A. G. Tilkian. Understanding Heart Sounds and Murmurs. Philadelphia: W. B. Sanders Company. 1984.

3. A. Leatham. Auscultation of the Heart and Phonocardiography. Churchill: Livingstone Company, London,1975.

4. Rick A. Nishimura. Aortic Valve Diseas. Circulation. 2002;106:770-772.

5. C.M. Otto, B.K. Lind, D.W. Kitzman. Association of aortic-valve sclerosis with cardiovascular mortality and morbidity in the elderly. N Engl J Med. 1999; (3):142-7.

6. R. Rosenhek, U. Klaar, M. Schemper. Mild and moderate aortic stenosis, Natural history and risk stratification by echocardiography. Eur Heart. 2004;25(3):199-205.

7. Raphael Rosenhek. Natural History of Very Sever Aortic Stenosis. Circulation. 2010;121:151-156.

8. A. A. Sepehri, A. Gharehbaghi, T. Dutoit, A. Kocharian, A. Kiani. A novel method for pediatric heart sound segmentation without using the ECG. Comput. Meth. Prog. Biomed.. 2010;99:43–48.

9. A. Gharehbaghi, T. Dutoit, A. Sepehri, P. Hult, P. Ask. An automatic tool for pediatric heart sounds segmentation. Proc. Computing in Cardiology. 2011;38:37–40.

10. X. Zhang, L.-G. Durand, L. Senhadji, H. C. Lee, J. L. Coatrieux. Time-frequency scaling transformation of the phonocardiogram based of the matching pursuit method. IEEE Trans. Biomed. Eng. 1998;44:972–979.

11. C. Cortes, V. Vapnik. Support vector networks. Mach. Learn.. 1995;20:273–297.

12. V. Kecman. Learning and Soft Computing: Support Vector Machine. Neural Networks and Fuzzy Logic Models. Cambridge, MA: MIT Press, 2002.

13. K. Dosik, M. E. Tavel. Assessment of severity of aortic stenosis through time-frequency analysis of murmurs. Chest. 2003;124:1638–1644.

14. A. Voss, A. Mix, T. Hunber. Diagnosing aortic valve stenosis by parameter extraction of heart sound signals. Ann. Biomed. Eng. 2005;33:951–959.

15. M. El-Segaier, O. Lilja, S. Lukkarinen, L. Sörnmo, R. Sepponen, E. Pesonen. Computer-based detection and analysis of heart sound and murmur. Ann. Biomed. Eng. 2005;33:937–942.

16. I. Maglogiannis, E. Loukis, E. Zafiropoulos, A. Stasis. Support vectors machine-based identification of heart valve diseases using heart sounds. Comput. Meth. Prog. Biomed. 2009;95:47–61.

17. A. A. Sepehri, J. Hancq, T. Dutoit, A. Gharehbaghi, A. Kocharian, A. Kiani. Computerized screening of children congenital heart diseases. Comput. Meth. Prog. Biomed. 2008;92:186–192.

18. C. G. DeGroff, S. Bhatikar, J. Hertzberg, R. Shandas, L. Valdes-Cruz, R. L. Mahajan. Artificial neural network–based method of screening heart murmurs in children. Circulation. 2001;103:2711–2716.

19. Sanjay R. Bhatikar, Curt DeGroff, Roop L. Mahajan. A classifier based on the artificial neural network approach for cardiologic auscultation in pediatrics. Artificial Intelligence in Medicine. 2005;33: 251-260.

20. R. K. Sinha, Y. Aggarwal, B. N. Das. Back propagation artificial neural network classifier to detect changes in heart sound due to mitral valve regurgitation. J. Med. Syst. 2007;31:205-209.

21. C. N. Gupta, R. Palaniappan, S. Swaminathan, S. M. Krishnan. Neural network classification of homomorphic segmented heart sounds. Appl. Soft Eng 2007;7:286–297.

22. R. K. Sinha, Y. Aggarwal, B. N. Das. Back propagation artificial neural network classifier to detect changes in heart sound due to mitral valve regurgitation. J. Med. Sys. 2007;31:205–209.

23. W. C. Kao, C. C. Wei. Automatic phonocardiograph signal analysis for detecting heart valve disorders. Expert Systems with Applications. 2011;38:6458–6468.

Assessment of the Microcirculation Using Combined Model Based Diffuse Reflectance Spectroscopy and Laser Doppler Flowmetry

H. Jonasson[1], I. Fredriksson[1,2], M. Larsson[1], and T. Strömberg[1]

[1] Department of Biomedical Engineering, Linköping University, Linköping, Sweden
[2] Perimed AB, Datavägen 9a, 175 26 Järfälla-Stockholm, Sweden

Abstract — By using a combined inverse model for diffuse reflectance spectroscopy (DRS) and laser Doppler flowmetry (LDF) the tissue fraction of red blood cells (RBCs), their oxygenation and speed-resolved perfusion are estimated in absolute units. DRS spectra (450 to 850 nm) are measured at two source-detector distances; 0.4 and 1.2 mm. LDF spectra are measured at 1.2 mm, integrated in the same fiber-optic probe. Inverse Monte Carlo technique and an adaptive tissue model is used to quantify the microcirculatory parameters. Measurements were done during venous occlusion of the tissue. The model fitting yields a good spectral fit for the two DRS spectra and the LDF spectrum. The physiological responses regarding for example which speed regions respond to provocations follows a priori expectations. The combined model gives quantitative measures of RBC tissue fraction, oxygenation and speed resolved perfusion from the same sampling volume which gives new opportunities to interpret data.

Keywords— **diffuse reflectance spectroscopy, laser Doppler flowmetry, modeling, Monte Carlo simulations.**

I. Introduction

The microcirculation is responsible for delivering oxygen to the local tissue through the red blood cells (RBC). Different tissues have different needs and both oxygenation and blood flow are important to consider when assessing the microcirculation.

We present an inverse Monte Carlo method using a multilayered tissue model for analyzing both diffuse reflectance spectroscopy (DRS) and laser Doppler flowmetry (LDF) data. The output from the model is an assessment of the RBC tissue fraction in mass percentage (%) and their oxygenation in (%), together with perfusion in absolute units (% RBC x mm/s) resolved in three speed regions (0-1, 1-10 and above 10 mm/s).

II. Method

The model of the skin in the wavelength range of 450 to 850 nm consisted of three layers; one epidermis layer with variable thickness and two dermis layers where the upper dermis layer had a fixed thickness of 0.5 mm and the lower had an infinite thickness. The epidermis layer was bloodless while the two dermis layers had different amount of blood with variable speed and equal oxygen saturation. The scattering and absorption properties in the model were described by ten parameters [1], including compensation for the vessel-packaging effect [2, 3]. The three-layer model and the ten parameters are a further development of the model described in [4]. The blood in the dermis layers had a speed distribution that was given by ten parameters. In total the model consisted of 20 parameters (1 epidermal thickness, 1 melanin exponent, 1 melanin fraction, 3 scattering parameters, 2 blood tissue fractions, 1 oxygen saturation, 1 mean vessel diameter and 10 speeds).

By finding a model that fits both DRS and LDF spectra, the inverse problem is solved and the 20 parameters are obtained. This is done in a non-linear optimization algorithm where simulated data are fitted to measured data, initially for DRS and then for LDF. Included in the error function are the intensity difference between simulated and measured data, with an emphasis of wavelengths in the hemoglobin absorption peak region (520 to 600 nm). The error function also penalize non physiological values on the parameters and deviations from unity on the intensity relaxation factor [5].

Measurements were done on one healthy male, age 32, with Caucasian skin. A 5 minutes venous occlusion of the forearm was conducted where the fiber-optic probe was placed in a probe holder and placed on the volar side of the lower forearm. The occlusion preceded by a 5 minutes baseline phase and followed by a 5 minutes reperfusion phase.

DRS spectra were collected at two detector distances (0.4 and 1.2 mm) from a broadband white light source (AvaLight-HAL-S, Avantes BV, The Netherlands) by a multi-channel spectroscope (AvaSpec 2048-5 RM, Avantes BV, The Netherlands). LDF spectra were collected from six detecting fibers placed 1.2 mm from a laser light source (780 nm) and connected to a single detector in a modified Periflux 5000 system (Perimed AB, Järfälla, Sweden).

Calibration of DRS spectra was done in three steps; dark subtraction, white normalization and relative calibration between the two channels, according to Fredriksson et al [4].

H. Mindedal and M. Persson (eds.), *16th Nordic-Baltic Conference on Biomedical Engineering*,
IFMBE Proceedings 48, DOI: 10.1007/978-3-319-12967-9_14

III. RESULTS

The RBC tissue fraction, the oxygen saturation and the speed resolved perfusion during the venous occlusion are depicted as a function of time in Fig. 1. The occlusion started at 5 minutes and lasted for another 5 minutes. The oxygenation drops during the venous occlusion while the RBC tissue fraction increases. The drop in total perfusion during the occlusion is caused by the decrease in perfusion for velocities above 10 mm/s and between 1 and 10 mm/s. The perfusion for speed below 1 mm/s is unaffected during occlusion.

The average relative root mean square (RMS) error of the spectral fit for the two DRS distances were 1.0 % for the 0.4 mm distance and 3.7 % for the 1.2 mm distance. For the LDF spectrum the average RMS was 4.7 % (in the logarithmic scale). An example of the fitting for the two DRS channels is given in Fig. 2. The fitting for the one LDF channel is given in Fig. 3.

IV. DISCUSSION

By integrating DRS and LDF not only in the same probe, but also in the model and the spectral fitting gives new opportunities. The fitting of the LDF spectra benefits from the DRS spectral fitting which gives a more accurate estimation of tissue scattering and the vessel packaging effect.

The adaptive three-layered tissue model is fast enough for real-time analysis which gives opportunities to monitor changes in RBC tissue fraction, oxygenation and perfusion while it happens, instead of having to do the analysis afterwards. Also, doing LDF and DRS measurements in the same vascular bed gives new opportunities to interpret data when having more information on the vascular status.

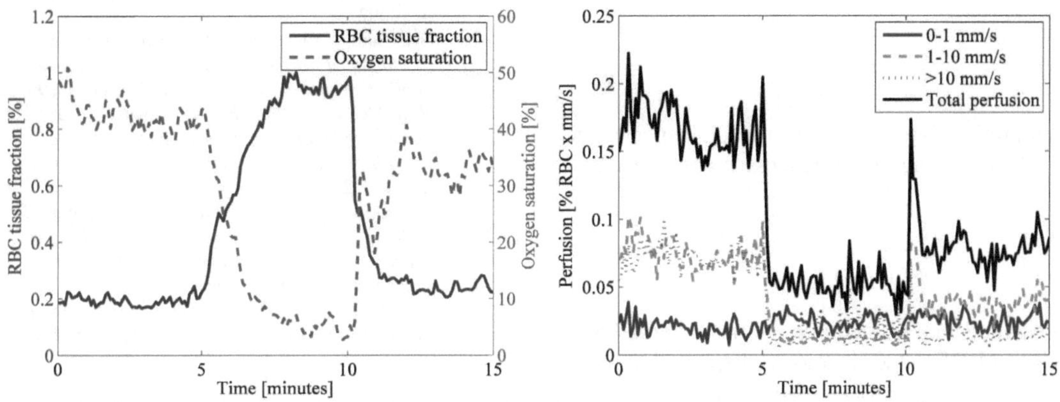

Fig. 1. Time-resolved RBC tissue fraction and oxygenation during venous occlusion (left) and speed resolved perfusion (right).

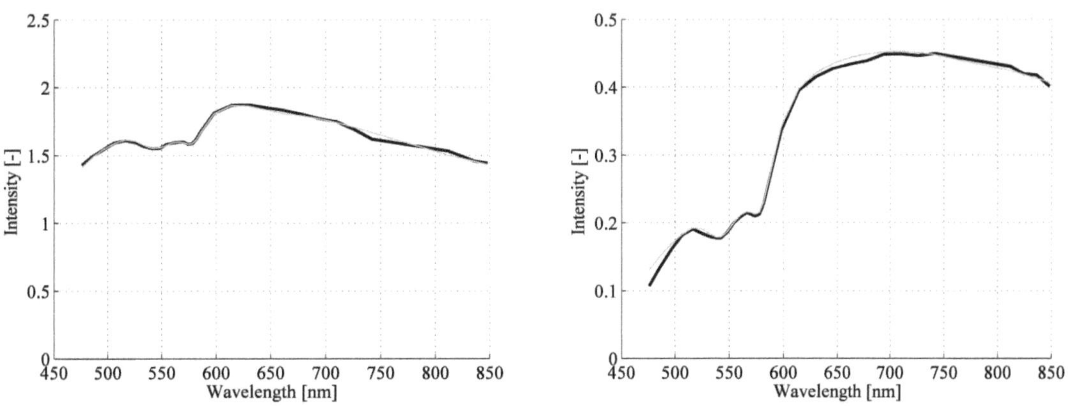

Fig. 2. An example of spectral fit of the DRS spectra at 0.4 mm (left) and 1.2 mm (right) at the beginning of the measurement (black: measured, yellow: fitted).

The oxygenation, the RBC tissue fraction and the perfusion shows expected values during venous occlusion. There is an increase in RBC volume fraction while the perfusion is reduced. This is an indication of the accuracy of the method. Other measurements with provocations like local heating and systolic occlusions have been performed (not presented in this abstract) and shows the same expected results.

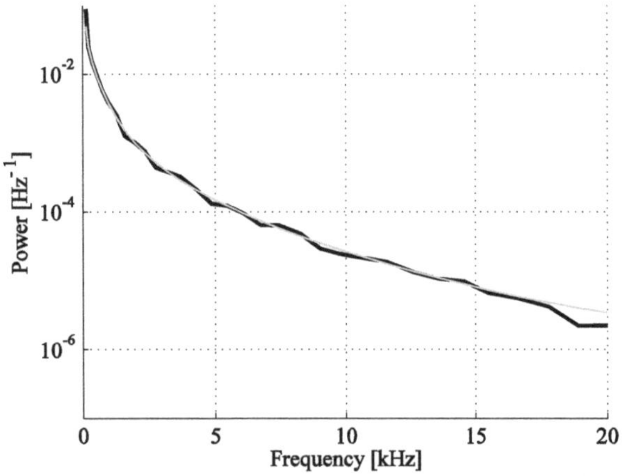

Fig. 3. An example of spectral fit for the LDF spectra at the beginning of the measurement (black: measured, yellow: fitted).

V. CONCLUSIONS

We have presented a combined inverse tissue model for analyzing DRS and LDF data. The model displayed a good spectral fit for both DRS and LDF while being capable of estimating physiological parameters in absolute units. Applied to tissue, the estimated RBC tissue fraction, oxygenation and speed resolved perfusion displayed an a priori expected behavior during physiological provocations.

ACKNOWLEDGMENT

This study was financed by VINNOVA and Perimed AB through the SamBIO research collaboration program between companies and academia within bioscience (D.no 2007-00149) and the Research&Grow program (D.no 2011-03074), and also by NovaMedTech, supported by the European Union Regional Development Fund.

REFERENCES

1. Strömberg, T., et al., *Microcirculation assessment using an individualized model for diffuse reflectance spectroscopy and conventional laser Doppler flowmetry.* J Biomed Opt, 2014. **19**(5): p. 57002.
2. Svaasand, L.O., et al., *Therapeutic response during pulsed laser treatment of port-wine stains: Dependence on vessel diameter and depth in dermis.* Lasers in Medical Science, 1995. **10**(4): p. 235-243.
3. Van Veen, R.L.P., W. Verkruysse, and H.J.C.M. Sterenborg, *Diffuse-reflectance spectroscopy from 500 to 1060 nm by correction for inhomogeneously distributed absorbers.* Optics Letters, 2002. **27**(4): p. 246-248.
4. Fredriksson, I., M. Larsson, and T. Strömberg, *Inverse Monte Carlo method in a multilayered tissue model for diffuse reflectance spectroscopy.* Journal of Biomedical Optics, 2012. **17**(4).
5. Fredriksson, I., et al., *Inverse Monte Carlo in a multilayered tissue model: merging diffuse reflectance spectroscopy and laser Doppler flowmetry.* J Biomed Opt, 2013. **18**(12): p. 127004.

Selection of EEG Frequency Bands for Detection of Depression

K. Kalev and M. Bachmann

Department of Biomedical Engineering, Technomedicum, Tallinn University of Technology, Tallinn, Estonia

Abstract— **Major depression affects more than 18 million people in the United States every year. Early diagnosis is essential for appropriate treatment of depression by promoting remission and by preventing relapses. With the aim of improving the early diagnosis of depression we analyzed the power spectrum of electroencephalographic (EEG) signal of female depressive subjects (17) and female control subjects (17). Earlier studies have found significant differences in depressive patient EEG power spectrum compared to healthy control subjects. These studies have used traditional EEG frequency bands, which have not been selected for diagnostic purposes of depression. In current study we evaluated EEG relative power in different frequency bands from frequency range 1 - 40 Hz in channel P3 - Cz to find the frequency bands best differentiating depressive subjects from controls. In addition, the linear discriminant analysis through leave-one-out cross-validation was applied. Best results, differentiating depressive subjects from controls, were obtained from frequency band 5 - 7 Hz which is the narrow subset of traditional theta frequency band (4 - 8 Hz). Depression specific relative beta and relative gamma frequency power yielded also smaller classification errors than traditional beta and gamma frequency bands. However, the classification error of traditional alpha band was large and the systematic analyze of different frequency band powers in alpha range did not improve the classification error. In future, it would be interesting to analyze the modified relative band power excluding the alpha frequencies.**

Keywords— **EEG, depression, power spectrum.**

I. INTRODUCTION

It has been established that more human suffering has resulted from depression than from any other single disease affecting mankind [1]. Major depression affects more than 18 million people in the United States (6% of the population) every year while involving the lives of about 200 million people. These figures are comparable to other developed countries [2].

Early diagnosis is essential for appropriate treatment of depression. This would promote remission, prevent relapse and reduce the economic burden of the disease [3]. However, the diagnosis of depression is still based on evaluation of the intensity of subjective symptoms by psychiatrists. Currently no objective indicator in clinical practice exists. Therefore, it would be beneficial to have objective indicators for diagnosing depression.

Electroencephalogram (EEG) is a method used to measure brain bioelectrical activity. It is easily available, cost effective and can provide objective information about changes in brain physiology specific in depression. Several studies have shown that EEG power spectrum analysis provides promising results for differentiating depression [4–7]. However, these studies use traditional EEG frequency bands, which vary slightly from study to study: 0.5-4 Hz (delta), 4-8 Hz (theta), 8-10 Hz (alpha1), 10-12 Hz (alpha 2), 12-20 Hz (beta 1), 20-30 Hz (beta 2), 30-40 Hz (gamma) [8, 9].

The results of earlier studies on depression indicate significant increase in absolute beta power (~13-30 Hz) in depression compared to controls [6]. The relative beta power (12.5-25 Hz) has also been shown to increase in depression [5]. However, in the study by Omel'chenko *et al.*, statistically significant decrease in relative beta power (14-30 Hz) was presented [7].

Changes in depression reported in absolute or relative theta frequency band (~4-8 Hz) have been more consistent. Authors report mainly statistically significant increase in theta power [4,6,7]. With the exception by Knott *et al.*, reporting no difference in absolute or relative theta power between depressive and control subjects [5].

Other studies have found various statistically significant differences in depression compared to controls: significant increase in absolute or relative delta power [6,7], increase [4,10] and decrease [6,7] in alpha power. Knott et al. reported significant increase only in beta power [5]. While another study reports that all analyzed frequency band powers indicated significant increase in depression [4].

As the results of analyzing depression are contradictive, EEG is still not used as an objective method in the process of diagnosing mental disorders. The inconsistency is likely due to different recording protocols, group sizes, intra-individual differences etc. However, also the used frequency bands vary from study to study. The most consistent theta frequency band gives also the most consistent results. Nevertheless, as earlier studies used traditional frequency bands, chosen arbitrarily, depression specific selection of frequency bands could improve the diagnostic capability of EEG frequency band power.

For instance, Elgendi *et al.* systematically analyzed the

H. Mindedal and M. Persson (eds.), *16th Nordic-Baltic Conference on Biomedical Engineering*,
IFMBE Proceedings 48, DOI: 10.1007/978-3-319-12967-9_15

EEG relative power of different frequency bands in the frequency range 1-30 Hz to distinguish between Alzheimer patients and healthy controls [11]. Selected, Alzheimer disease specific frequency bands were found to have significantly better classification performance than traditional frequency bands.

In current study, we are going to expand the idea on EEG in depression by finding the frequency bands best differentiating depressive and healthy subjects. For this purpose we evaluate relative power in different frequency bands from frequency range 1- 40 Hz. In addition, the linear discriminant analysis through leave-one-out cross-validation will be applied.

II. Methods and Equipment

A. Subjects

The experiment was carried out on a group of patients with depressive disorder and a control group. As depression is more common in females [12], we explored only female subjects in this study. Both groups consisted of 17 subjects with a mean age of 39 years, with standard deviation 12 years. Subjects with nonpsychotic depressive disorder as defined by ICD-10 criteria and determined by 17-item Hamilton Depression Rating Scale (HAM-D) score higher than 14 were eligible.

The experiments were conducted with the understanding and written consent of each subject in accordance with the Declaration of Helsinki and approved by the Tallinn Medical Research Ethics Committee.

B. Experimental procedure and EEG recording equipment

The experimental procedure for a subject included continuous EEG recording during 30 minutes between time interval 9 a.m. to noon. The subjects were lying in a relaxed position, eyes closed and ears blocked during the experiments.

The EEG time series were recorded using 18 channels according to the international 10-20-electrode position classification system. The Cadwell Easy II EEG measurement equipment was used for recording raw EEG signals with a frequency band of 0.3-70 Hz, at a sampling frequency of 400 Hz. The EEG signals were further digitally filtered with fifth-order Butterworth filter between 1 and 40 Hz. The EEG analysis was performed in channel P3, as parietal region was previously proven effective while studying depression [13]. In addition, only the first 5 minutes of the EEG signal were analyzed with a perspective to find most effective frequency bands for short time segments. The EEG analysis was performed using MATLAB software.

C. EEG Analysis

The relative power was calculated in 1521 different frequency bands in the range of 1 to 40 Hz.

First, the absolute power of the recoded EEG data was calculated by means of Welch's averaged periodogram method. The signal was divided into overlapping segments (50% overlapping) with the length of 2,000 samples. Every segment was multiplied by Hamming window. The periodogram was calculated by applying fast Fourier transformation (FFT) to a windowed segment and time-averaged, which reduces the variance of the individual power measurements. The absolute power in certain frequency band was calculated as the integral of all power values over that frequency range.

Next, the relative power of a certain frequency band, rP_{fw}, was calculated as the ratio of the absolute power of that frequency band, aP_{fw}, to the total power of full frequency band 1-40 Hz, aP_{all}:

$$rP_{fw} = \frac{aP_{fw}}{aP_{all}}, \tag{1}$$

where $f \in [1,39]$ Hz indicates the frequency bandwidth start and $w \in [1, 39-(f-1)]$ Hz indicates the width of the current frequency bandwidth.

The relative power in channel P3 in all frequency bands was calculated for all depressive and control subjects.

Next, the linear separability index [11] was determined as

$$J_{fw} = \frac{|\mu c_{fw} - \mu d_{fw}|}{\sigma c_{fw} + \sigma d_{fw}}, \tag{2}$$

where μc_{fw} and μd_{fw} indicate the average rP_{fw} for control and depressive subjects for different frequency bands and σc_{fw} and σd_{fw} represent the standard deviation of rP_{fw} for control and depressive subjects for different frequency bands.

D. Statistical Analysis

Wilcoxon Rank-Sum statistical test was performed to evaluate the differences between depressive patients' and healthy control subject's relative power of various frequency bands. Low p-values indicate large difference in the median of relative power of depressive patients and healthy control subjects. *P*-values less than 0.05 were considered as significant.

The linear discriminant analysis (LDA) with leave-one-out cross-validation *(LOO)* was applied for each rP_{fw} separately to classify the depressive and control subjects.

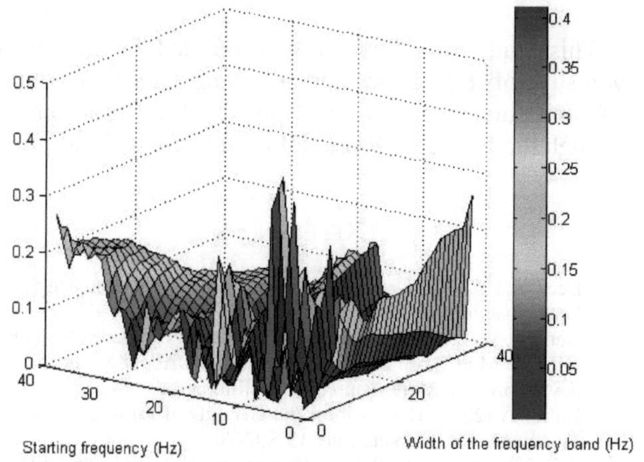

Fig. 1 *Linear separability index J (z-axis) calculated for different frequency bands. The x-axis represent the width of the analyzed frequency band. The y-axis represents the starting frequency of the analyzed frequency band.*

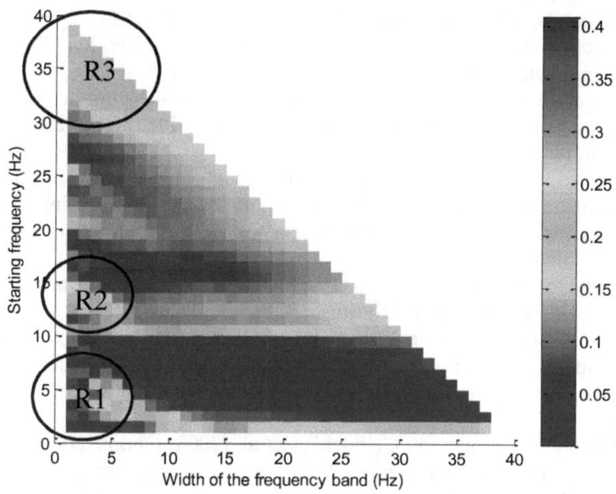

Fig. 1 *Colorscaled linear separability index J calculated for different frequency bands. The x-axis represent the width of the analyzed frequency band. The y-axis represents the starting frequency of the analyzed frequency band. Three regions with highest linear separability values in different tradiotional EEG frequency bands are marked with circles (R1, R2, R3).*

III. RESULTS AND DISCUSSION

Linear separation values for all frequency bands are presented in Fig 1. On Fig 2 the same results are presented with regions of high linear separability (R1, R2, R3).

Table 1 Best linear separability value J between depressive and healthy control subjects for all three regions which are depicted in Fig. 2

Region 1		Region 2		Region 3	
Frequency band (Hz)	J	Frequency band (Hz)	J	Frequency band (Hz)	J
5-7	0.4084	14-15	0.2525	39-40	0.2668
5-6	0.3883	13-15	0.2370	38-40	0.2424
4-7	0.3641	13-14	0.2147	35-36	0.2258
6-7	0.3534	13-16	0.2059	35-37	0.2214
4-6	0.2831	12-15	0.1979	38-39	0.2207
3-7	0.2479	...	...	...	...
4-8	0.2289	12-20	0.1135	30-40	0.1907

Table 2 P-values and LDA error for traditional frequency bands and selected frequency bands which had smallest LDA error value

	Traditional frequency band			Selected frequency band		
Measures	4-8 Hz	12-20 Hz	30-40 Hz	5-6 Hz	13-15 Hz	39-40 Hz
p-value	3.6×10^{-2}	8.4×10^{-1}	5.6×10^{-1}	7.2×10^{-3}	3.8×10^{-1}	3.5×10^{-1}
Error of LDA	37.7%	46.5%	49.9%	23.8%	35.7%	41.3%

Table 3 P-values and LDA error for frequency bands which had highest linear separability values

	Top frequency bands				
	5-7 Hz	5-6 Hz	4-7 Hz	6-7 Hz	4-6 Hz
p-value	7.2×10^{-3}	7.2×10^{-3}	1.8×10^{-2}	9.8×10^{-3}	7.3×10^{-2}
Error of LDA	28.2%	23.8%	31.0%	33.8%	26.7%

According to Fig 2, the highest linear separability values emerge in the theta frequency band (R1). While high linear separation values were also obtained in beta1 (R2) and gamma frequency band (R3). Interestingly, there is no peak in alpha frequency range, indicating that relative alpha power is not a good marker for differentiating depressive subjects from controls.

The best linear separability values in each region are presented in Table I. In region 1, the most effective frequency band is 5-7 Hz. It is the narrow subset of traditional theta frequency band (4-8 Hz). In region 2, 14-15 Hz gave the best result. In most EEG studies, the beta frequency is calculated in wide frequency ranges (12-20 Hz or even 13-30 Hz). Current results suggest that better results for depression could be achieved with more specific frequency ranges. In addition, the best frequency band in region 3 was 39-40

Hz. This belongs to gamma frequency band and is very often entirely neglected from EEG analysis. The last row in Table I indicates that optimized frequency bands have almost twice better linear separation values than traditional frequency bands.

In Table II, a comparison of traditional frequency bands (theta, beta, and gamma) and selected frequency bands, according to the lowest LDA error value, are presented. While comparing the classification error in traditional frequency bands, the smallest error appears in traditional theta frequency band - 37.7 %. After depression specific selection of theta frequency band, the classification error decreases to the value of 23.8 %. The improvement of LDA error is highest in theta frequency range. Almost as good improvement is gained in beta frequency band. However, as the LDA error for traditional beta frequency band was high (46.5 %), the selected beta frequency band classification error is still much higher (35.7 %) than in selected theta frequency range (23.8 %). Some improvement was noticed also for optimized gamma frequencies, while no improvement was revealed for alpha frequencies.

Considering the LDA error values of selected frequency bands (Table II), we can see that selected theta frequency is the best choice for differentiating depressive subjects from controls. In current study the separability index did not consider the direction of change in mean relative power between depressive and control subjects. In case the differences between depression and control group in theta and in beta relative power are in opposite directions, the combination of both could give even smaller LDA errors indicating better classification ability.

In Table III the top five highest linear separability values from all frequency ranges are presented. As all top five frequency bands belong to the theta frequency region, the effectiveness of theta band is demonstrated again.

The data set used in this paper is fairly small. Additional testing on larger datasets is needed to assure the effectiveness of the proposed approach. In addition, it would be interesting to calculate modified relative power excluding the alpha frequencies.

IV. CONCLUSIONS

The results indicate that depression specific selection of EEG frequency bands improves the differentiation of depressive subjects from controls in all EEG frequency bands except alpha. In addition, best results differentiating depressive subjects from controls, were obtained from theta frequency band.

ACKNOWLEDGMENT

This study was financially supported by the Estonian Ministry of Education and Research under institutional research financing IUT 19-2 and by the European Union through the European Regional Development Fund.

REFERENCES

1. Beck A.T (2009) Depression: Causes and Treatment. University of Pennsylvania Press, Philadelphia
2. Burgess W (2009) The Depression Answer Book Professional Answers to More Than 275 Critical Questions About Medication, Therapy, Support & More. Sourcebooks, Illinois
3. Halfin A (2007) Depression: The Benefits of Early and Appropriate Treatment, Am J Manag Care 13:S92-701
4. Grin-Yatsenko V.A, Baas I, Ponomarev V.A, Kropotov J.D (2009) EEG power spectra at early stages of depressive disorders. J Clin Neurophysiol 26:401-406
5. Knott V, Mahoney C, Kennedy S et al. (2001) EEG power, frequency, asymmetry and coherence in male depression. Psychiatry Res 106:123-140
6. Begić D, Popović-Knapić V, Grubišin J et al (2011) Quantitative electroencephalography in schizophrenia and depression. Psychiatr Danubina. 23: 355-362
7. Omel'chenko V.P (2002) Changes in the EEG-Rhythms in Endogenous Depressive Disorders and the Effect of Pharmacotherapy. Human Physiology 28: 275-281
8. Lee S, Jones S.R (2013) Distinguishing mechanisms of gamma frequency oscillations in human current source signals using a computational model of a laminar neocortical network. Front Hum Neurosci 7:869
9. Dehaene S, Christen Y (2011) Characterizing Consciousness: From Cognition to the Clinic? Springer Berlin Heidelberg
10. AH, Griffiths K, Felmingham KL et al. (2010) Disorder specificity despite comorbidity: resting EEG alpha asymmetry in major depressive disorder and post-traumatic stress disorder, Biol Psychol 85:350-4
11. Elgendi M, Vialatte F, Cichocki A et al. (2003) Optimization of EEG frequency bands for improved diagnosis of Alzheimer disease, Conf Proc IEEE Eng Med Biol Soc., 33rd Annual International Conference of the IEEE EMBS, Boston, Massachusetts USA, 2011, pp 6087-6091
12. NIMH at http://www.nimh.nih.gov/health/publications/the-numbers-count-mental-disorders-in-america/index.shtml
13. Hinrikus H, Suhhova A, Bachmann M et al. (2009) Electroencephalographic spectral asymmetry index for detection of depression, Med Biol Eng Comput 47:1291-9

Author: Kaia Kalev
Institute: Tallinn University of Technology, Technomedicum
Street: Ehitajate tee 5
City: Tallinn
Country: Estonia
Email: kaia.kalev@cb.ttu.ee

Fluorescence of Beta-2-microglobulin in the Spent Dialysate

S. Kalle, H. Kressa, R. Tanner, J. Holmar, and I. Fridolin

Department of Biomedical Engineering, Technomedicum, Tallinn University of Technology,
19086, Ehitajate tee 5, Tallinn, Estonia

Abstract— **The aim of the study was to determine if fluorescence chromatography can be used to measure modified beta-2-microgobulin (B2M) from the spent dialysate. Amyloid B2M is the main pathogenic component of dialysis-related amyloidosis. This component is in our sphere of interest being one of the fluorescent advanced glycation end products (AGE). AGEs are potential uremic toxins that can cause amyloidosis and cardiovascular problems in chronic kidney failure patients.**

Two haemodialysis patients with high levels of B2M were selected for this study. Their spent dialysate samples were collected 10 minutes after the start of the dialysis process and less hydrophilic compounds were concentrated using solid phase extraction (SPE) column. Sediment from the concentrate and spent dialysate were analysed with electrospray ionisation mass spectrometer (ESI-MS) MicrOTOF-Q II coupled to high pressure liquid chromatography (HPLC) Dionex UltiMate 3000 RS. The sediment was analysed with Poroshell 120 EC-C18 column and spent dialysate with Kinetex C18 100A column. MagTran was used to interpret mass spectra.

Brown coloured fluorescent sediment of the concentrate was identified as amyloid B2M on the basis of MS and fluorescence spectra. AGE modified B2M was also found from spent dialysate. However the fluorescence intensity was very low compared to overall fluorescence of spent dialysate.

In summary, the study revealed that the fluorescence of AGE modified B2M is possible to detect in spent dialysate. However, the measuring system needs high selectivity and sensitivity for detection due to low contribution of AGE modified B2M to overall fluorescence.

Keywords— **Advanced glycation, beta-2-microgobulin, fluorescence, mass spectrum, dialysate**

I. INTRODUCTION

Advanced glycation end products (AGEs) are a heterogeneous group of molecules that accumulate in uremic patients' plasma [1]. AGEs are formed when a carbonyl of a reducing sugar condenses with a reactive amino group in target protein [2]. Free AGEs probably originate mainly from the glycation of proteins and from food. AGEs accumulate as the result of decreased excretion and increased formation from oxidative stress. Accumulation of AGEs may be linked to the increased risk of cardiovascular disease in patients with renal failure [3]. Also, AGEs can predict diabetic microvascular complications [4].

To date, various AGEs have been identified e.g. N(6)-carboxymethyl lysine – the most used marker for AGEs in food analysis [5]. Our interest has captured beta-2-microglobulin (B2M) which can be advanced glycated. B2M is a polypeptide with a molecular weight of 11.8 kDa. It is present on the surface of most nucleated cells and in biological fluids [6]. The normal serum B2M concentration is 1.5–3 mg/l, and its average normal production rate has been estimated to be 2.4 mg/kg/day [7]. In patients with chronic kidney failure, B2M concentration increases as the glomerular filtration rate decreases. The accumulation of AGE modified B2M can lead to different health complications like dialysis-related amyloidosis [2,8].

AGEs are classified as being potential uremic toxins [9], therefore removal of AGEs can be used as an indicator of the dialysis therapy. B2M could be a marker to describe the elimination of middle molecular weight solutes [2] because levels of B2M are not affected by food intake [2]. As a group, AGEs are difficult to measure, but individual compounds have been analyzed. There is no universal method for AGE measurements. B2M is mostly measured via B2M human enzyme-linked immunosorbent assay (ELISA) kit or high pressure liquid chromatography (HPLC). In a health care setting these methods are expensive and the results cannot be used to modify an ongoing dialysis process. An online measuring method would be advantageous. Fluorescence detector is one device that could be used in online measurements.

The aim of the study was to determine if fluorescence chromatography can be used to measure AGE modified B2M from spent dialysate.

II. MATERIALS AND METHOD

This study was performed after approval of the protocol by the Regional Ethical Review Board, Estonia. An informed consent was obtained from all participating patients.

Two haemodialysis patients with high levels of B2M were selected for this study. Their spent dialysate samples were collected from the drain tube of the dialysis machine 10 minutes after the start of the dialysis session. Less hydrophilic compounds were concentrated from spent dialysate using C18 solid phase extraction (SPE) column. The concentration consisted of 5 phases: (1) column was washed

H. Mindedal and M. Persson (eds.), *16th Nordic-Baltic Conference on Biomedical Engineering*,
IFMBE Proceedings 48, DOI: 10.1007/978-3-319-12967-9_16

with 120 ml of methanol (MeOH); (2) MeOH was washed away with 120 ml milliQ water; (3) 200 ml of dialysate was added; (4) 120 ml of milliQ water was used to wash away salts from sorbent; (5) 45 ml MeOH was added and the concentrate was collected. After that the concentrate was put to +6°C storage and sedimentation occured. Sediment of the concentrate was dissolved in 3 ml of 4:1 urea (6 M): NaCl (9%) solvent and centrifuged before analysis.

The sediment and original dialysate were analysed with electrospray ionisation mass spectrometer MicrOTOF-Q II (Bruker Daltoniks, Germany) coupled to HPLC Dionex UltiMate 3000 RS (Dionex, USA). For sediment's fluorescence measurements spectrofluorometer RF-5301PC (Shimadzu, Japan) and for original spent dialysate fluorescence detector RF 2000 (Dionex, USA) was used. Analysis was performed in the excitation (EX) range of 220 – 480 nm and emission (EM) 240 – 500 nm. The sediment was analysed with Poroshell 120 EC-C18 column (Agilent Technologies, USA) and spent dialysate with Kinetex C18 100A column (Phenomenex, USA). Mass spectra interpretations were done with MagTran (Amgen Inc, USA) software.

Chromatographic measurements were conducted with acetonitrile (Romil, UK) and 0.05M ammonium formate (pH 4.25). Samples were adjusted to pH 4.25 and centrifuged for 10 minutes before injection.

III. Results

The fluorescence chromatograph of sediment showed several peaks in the long wavelength region (EX352 nm/ EM460 nm). One of the peaks (Fig. 1) was found to coincide with B2M on the MS chromatograph.

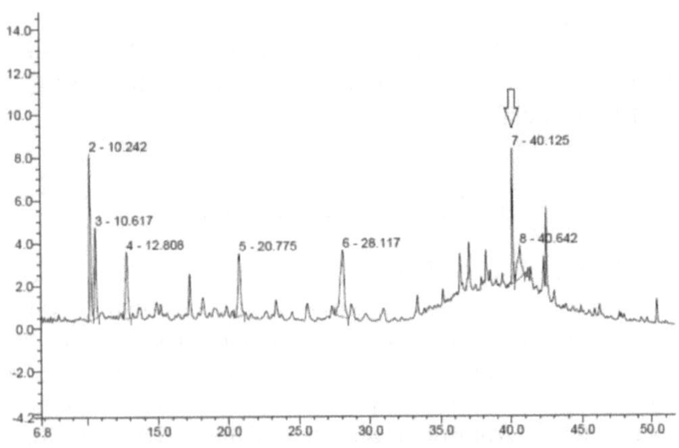

Fig. 1 Fluorescence chromatograph of the sediment. Peak number 7 is the peak coinciding with B2M in the MS chromatograph. EX 352 nm/ EM 460 nm (x-axis is time in minutes, y-axis is fluorescence intensity in mV).

Mass spectrum (MS) of the sediment showed ions with multible charges. The molecule mass was found to be 11772 Da (Fig. 2). Also fluorescence spectra of the sediment and amyloid B2M were compared (Fig. 3).

Analysis of the spent dialysate was done under the same conditions and a polypeptide presence was found (Fig. 4). The data from MS shows several masses close to B2M (Fig. 5).

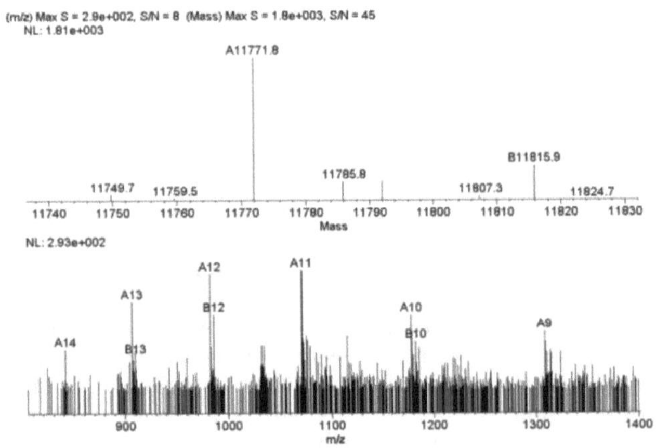

Fig. 2 Sediment's MS spectrum interpertated with MagTran software showing masses belonging to B2M. Charges of ions (down) and deconvulted masses (up).

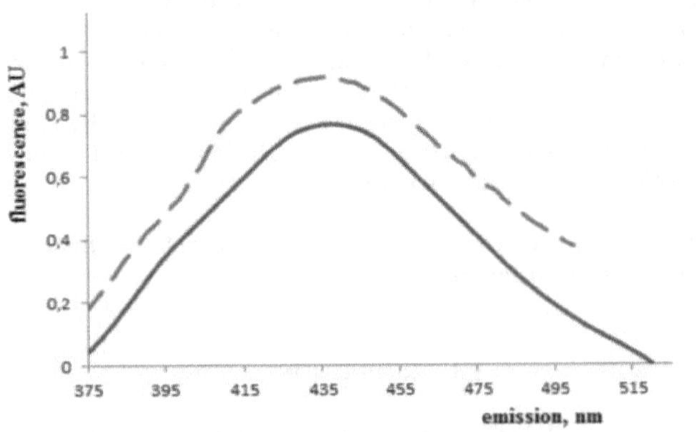

Fig. 3 Sediment's fluorescence spectrum (dashed line) compared with the spectrum of amyloid B2M (solid line) [8]

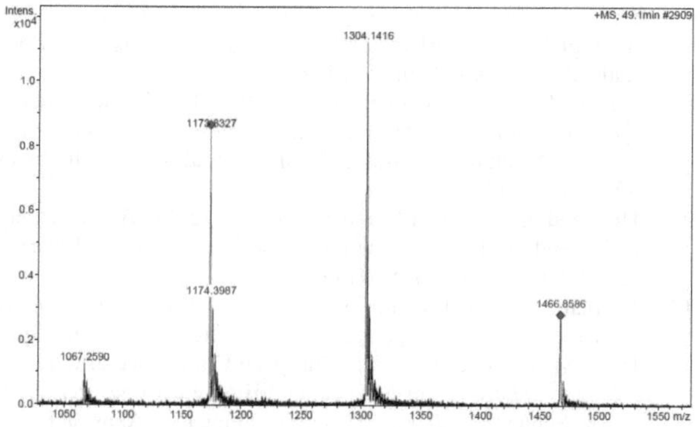

Fig 4 MS of spent dialysate showing a presence of protein.

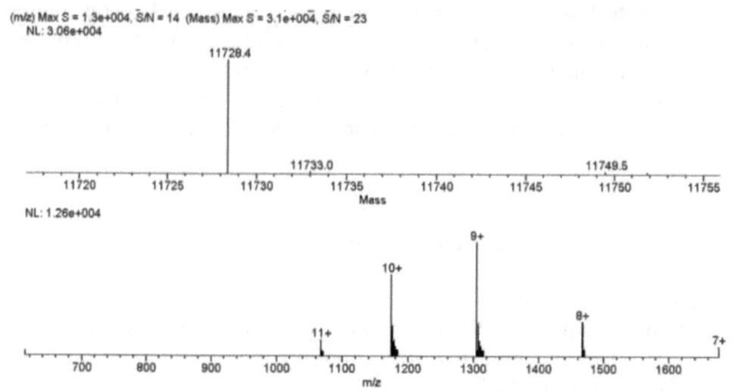

Fig. 5 MS of spent dialysate showing masses belonging to B2M Possible charges of ions (down) and molecule masses (up).

The chromatograph of spent dialysate had many fluorescent peaks in different wavelengths, but we were able to identify B2M peak (Fig. 6).

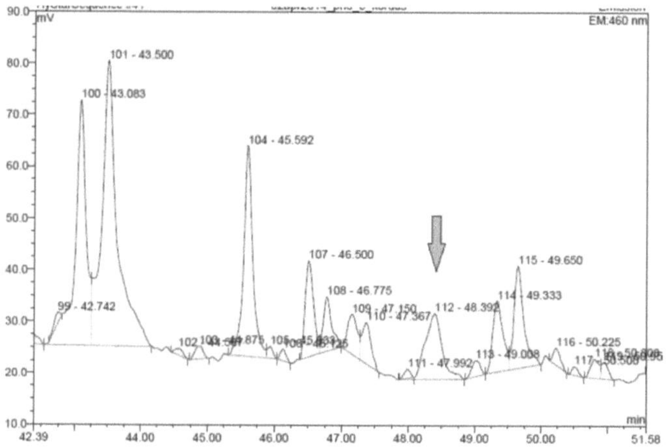

Fig. 6 Fluorescence chromatograph of spent dialysate. Under number 112 is B2M peak (EX 352 nm/ EM 460 nm).

IV. Disccusion

For the present study fluorescence detector was selected because it is easy and cost effective method that could be added to an online monitoring system. Also a good correlation (R=0.95) between B2M concentrations and fluorescence intensity of the spent dialysate was found earlier by our research group [10]. Our latest study, including 10 patient's data, showed again a relatively good R value of 0.74 (unpublished data).

Sedimentation occurred after concentrating two patients' spent dialysate samples with a SPE column. Analysis of the sediment showed that it had fluorescence characteristic to amyloid B2M (figure 3). Unmodified B2M does not have fluorescence but AGE modified compounds have specific fluorescence [8]. Many AGEs have an EX maximum approximately 370 nm and EM maximum around 445 nm [11] but the data varies. For an example, crossline's EM maximum is around 460 nm [12]. Our measurements were done near the maximum EM range of AGEs as seen from figure 3.

Both sediment's and dialysate's corresponding mass spectra showed typical ion masses of a polypeptide. The slight difference in masses can be explained by different degrees of modifications of the protein in amyloid sediment compared with the protein in original dialysate. Finding fluorescent modified B2M from the sediment let us assume that B2M can be measured from the unconcentrated dialysate. From spent dialysate's MS (Fig. 4) molecular masses belonging to B2M (11728 Da) were found (Fig. 5). However, the fluorescence intensity of modified B2M is low (0.04%) compared to overall fluorescence. This indicates that for AGE modified B2M detection a measuring system with high selectivity and sensitivity is necessary.

V. Conclusions

Brown coloured sediment and spent dialysate were analysed with HPLC-MS and fluorescence spectroscopy. The fluorescent sediment of the concentrate was identified as amyloid B2M on the basis of MS and fluorescence spectra. AGE-modified B2M was also measured from spent dialysate. In summary, the study revealed, that the fluorescence of amyloid B2M is possible to detect in spent dialysate, but the contribution to the overall fluorescence in very low. Thus, the measuring system needs high selectivity and sensitivity for detection.

Acknowledgment

The study was partly supported by the Estonian Science Foundation Grant No 8621, by the European Union through the European Regional Development Fund and institutional research funding IUT19-02.

Conflict of Interest

The authors declare that they have no conflict of interest.

References

1. Raj DS, Choudhury D, Welbourne TC, Levi M (2000) Advanced glycation end products: a Nephrologist's perspective. Am J Kidney Dis 35(3):365-80.
2. Vanholder R, et al. (2008) Do we need new indicators of dialysis adequacy based on middle-molecule removal? Nature Clinical Practice Nephrology 4: 174-175.
3. Agalou S1, Ahmed N, Dawnay A et al. (2003) Removal of advanced glycation end products in clinical renal failure by peritoneal dialysis and haemodialysis. Biochem Soc Trans. 31(6):1394-6.
4. Sampathkumar R, Balasubramanyam M, Rema M. et al. (2005) A novel advanced glycation index and its association with diabetes and mi-croangiopathy. Metabolism, 54, 1002-1007.
5. Semba RD, Nicklett EJ, Ferrucci L (2010). Does accumulation of advanced glycation end products contribute to the aging phenotype?. J Gerontol 65A (9): 963–975.
6. Vincent C, Revillard J. (1981) b2M and HLA-related glycoproteins in human urine and serum. Contr Nephrol 26:66–88.
7. Floege J et al. (1991) Clearance and synthesis rates of beta 2-microglobulin in patients undergoing hemodialysis and in normal subjects. J Lab Clin Med 118:153–165.
8. Miyata T, Oda O, Inagi R et al. (1993) Beta 2-Microglobulin modified with advanced glycation end products is a major component of hemodialysis-associated amyloidosis. J Clin Invest 92(3): 1243–1252.
9. Heidland A, Sebekova K and Schinzel R. (2001) Advanced glycation end products and the progressive course of renal disease. Am. J. Kidney Dis. 34, S100–S106
10. Holmar J, Arund, J; Uhlin, F et al. (2011). Beta2-microglobulin Measurements in the Spent Dialysate Using Fluorescence Spectra. IFMBE Proceedings: 5th European Conference of the International Federation for Medical and Biological Engineering, 14 – 18 September 2011, Budapest, Hungary. Ákos Jobbágy. Springer Heidelberg, 1035 - 1038.
11. Maza MP, Garrido F, Escalante N et al. (2012) Fluorescent advanced glycation end-products (ages) detected by spectrophotofluorimetry, as a screening tool to detect diabetic microvascular complications. J Diabetes Mellitus 2(2): 221 – 226.
12. Obayashi H, Nakano K, Shigeta H et al. (1996) Formation of Crossline as a Fluorescent Advanced Glycation End Productin Vitroandin Vivo. Biochem Biophys Res Commun 226(1): 37 – 41.

Address of the corresponding author:

Author: Sigrid Kalle
Institute: Department of Biomedical Engineering, Technomedicum, Tallinn University of Technology
Street: Ehitajate tee 5
City: 19086 Tallinn
Country: Estonia
Email: sigrid.kalle@ttu.ee

Application of 3D Scanner for Estimation of Chest Movement in Scoliotic Patients

A. Katashev[1], K. Romberg[2], A. Danielsson[3], and H. Saraste[4]

[1] Riga Technical University, Institute of Biomedical Engineering and Nanotechnology, Riga, Latvia
[2] Sahlgrenska University Hospital, Physiotherapy Department, Göteborg, Sweden
[3] Sahlgrenska Academy at the Göteborg University, Göteborg, Sweden
[4] Karolinska Institutet, Department of Molecular Medicine and Surgery, Stockholm, Sweden

Abstract— **Severe spine deformities cause restrictive respiratory insufficiency due to asymmetry of the chest wall and restriction of corresponding breathing movements. An important goal of the treatment is to restore flexibility of the chest wall to improve the breathing movement pattern. To monitor the progress of the chest wall deformity and its treatment dependent changes, a tool for assessment of the chest wall movement is required. Several non-radiological techniques have been proposed, such as marker based motion capture systems and photogrammetric methods. The current project describes an application of 3D optical system, based on 4 Artec® scanners, for monitoring and mapping of the asymmetry of the chest wall movements. Captured data were used to quantify the breathing related volume differences in inspiration and expiration separately on the left and right body side and visualize their locations. Asymmetry index was calculated as a difference in volume changes between right and left body sides, divided by the total volume change. Although the advantages of the method - short time for data capturing and application - are of importance in pediatric use and make this method attractive, the inaccuracy of the volume change estimation, being more then ±13% in the best case, is unacceptable and need to be reduced.**

Keywords: **breathing movements, chest wall, trunk asymmetry, scoliosis, 3D scanner.**

I. Introduction

Scoliosis and other spine deformities cause restrictive respiratory insufficiency due to asymmetry of the chest wall and restriction of chest wall movements [1]. Young children are most vulnerable, since moderate to major deformities deteriorate proper development and growth of their lungs resulting in restrictive ventilation deficiency and insufficient gas exchange [2, 3]. In pediatric spine deformities, the main treatment goal is to prevent the progression of both spine and chest wall deformities as well as to restore flexibility of the chest wall to improve the breathing movement pattern and pulmonary function. Corresponding tests used to evaluate treatment outcome include several pulmonary tests, such as vital capacity and forced vital capacity [4, 5]. These methods cannot present a possible asymmetry of the restriction location. This might be done, for instance, by means of radioactive aerosol ventilation scintigraphy [6], however with a noticeable radiation dose to patient. The radiation is also a problem for computed tomography [7], too, but not for dynamic magnetic resonance imaging [3]. These two methods allow detailed 3D reconstruction of lung volume change and could be used as a "gold standard", but high costs and restrictions in patient positioning still would be a limitation.

Other possible tools to assess and document deformity and its treatment dependent changes are those based on motion capture technology, such as ultrasonography [8] and optical techniques [9]. Unfortunately, these methods often need a large number of markers to identify anatomical points and to trace chest wall movements. For instance, more then 80 markers had been used for one of the most accurate measurements [10], which makes the procedure too time consuming and poorly tolerated by pediatric patients with severe deformity and/or young age. The marker free approaches include photogrammetry [11] and use of 3D depth camera [12]. The present research explores the possibility to use commercially available optical 3D scanners for dynamic acquisition of human trunk shape for evaluation of breathing volume changes and asymmetry of the chest wall.

II. Materials and Methods

A. Experimental setup and data acquisition

Four optical 3D scanners Artec3D® (Artec Group, San Diego, CA) were used to digitize patient 3D shape. Scanners were placed in the corners of a square area with the diagonal of 3 meters (Fig.1) and the height of approximately 1.2 meters from the floor. Thus, the effective "volume of view" of each scanner was approximately 90×76×50 cm (height × width × depth), allowing capture of 3D data of an adult patient trunk from the vertex to upper pelvis. The maximal rate of data capture was 15 frames per second.

The current project is a part of a larger study on respiratory function and quality of life of scoliosis patients, who had been treated for an early onset spine deformity at Sahlgrenska University Hospital. For the data acquisition, patients were asked to perform 3 deep breath cycles with

H. Mindedal and M. Persson (eds.), *16th Nordic-Baltic Conference on Biomedical Engineering*,
IFMBE Proceedings 48, DOI: 10.1007/978-3-319-12967-9_17

hands placed on the back of their heads. The posture was chosen to avoid the problem with disturbing arm volumes.

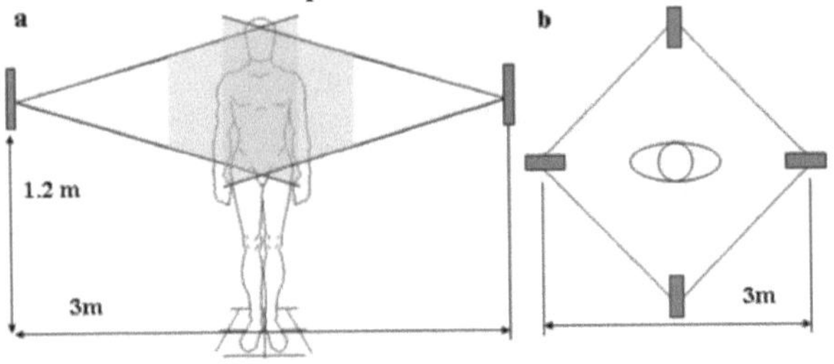

Fig. 1 Alignment of 4 ARTEC® scanners: a) front view, b) top view

Two data sets, one with, and one without markers, were captured. The markers were placed (Fig. 2) on the top of the sternum (S), on the spinous process of the seventh cervical (C7) and the fifth lumbar (L) vertebrae, and at the midline, about 2 cm superior to the pubic joint (P).

Although totally group of 45 adult patients was involved in the study, the present case study report is based on the typical data of the single patient. Results, obtained with other patients' data were similar.

B. Data processing and volume calculations

Pre-processing of the data was performed using commercial *Artec Studio®* software. It included manual removal of out-of object noise point debris and export of the scans into ".obj" format. Separate 3D objects, captured by each scanner, were aligned frame-by-frame using a custom-made software, which was more time saving than the one of the original capture software.

Patient trunk position was identified, using the six anatomical landmarks: S, C7, L, P and armpit marks A1 and A2 (Fig.2). For the non-marker data acquisition, the identification of the points on the image imitated the procedure of the marker placement on the patient: point S was marked in the jugular notch, C7 on the cervico-thoracic prominence, L5 as a point at the middle of the spine, between the right and left projection of iliac crest, and point P at the middle of the line connecting belly button and pubic joint. Right and left armpits, correspondingly, were marked at the proximal end of the triceps (A1, A2). Selection of all points was made on-screen by mouse clicking.

Anatomical landmarks were selected using expiration frame. For each selected anatomical landmark, the region of interest (ROI) was defined (Fig. 2). These ROI, that include neighboring points for corresponding landmark, were used for automatic alignment of inspiration and expiration frame. Such an alignment was made to exclude the influence of the posture sway, especially noticeable at deep breathing [13]. The alignment was made by iterative algorithm, minimizing closest point distances between landmark's ROI and corresponding frame [14]. In addition, the landmarks at the inspiration frame were selected manually, as well.

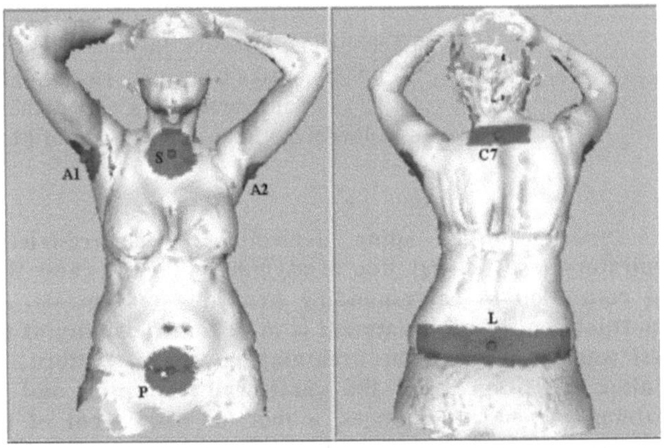

Fig. 2 Anatomical landmarks and their neighboring regions

Expiration and inspiration frames were aligned to match the spatial position of the landmarks L, direction of line segments L – C7 and planes P – L – C7 (fig.3a). The core trunk of the patient was defined by the outline of patient's 3D shape and three planes: the plane, going through line L – P perpendicularly to the plane P – L – C7 and planes C7 – A1 S and C7 – A2 – S (fig.3a,b).

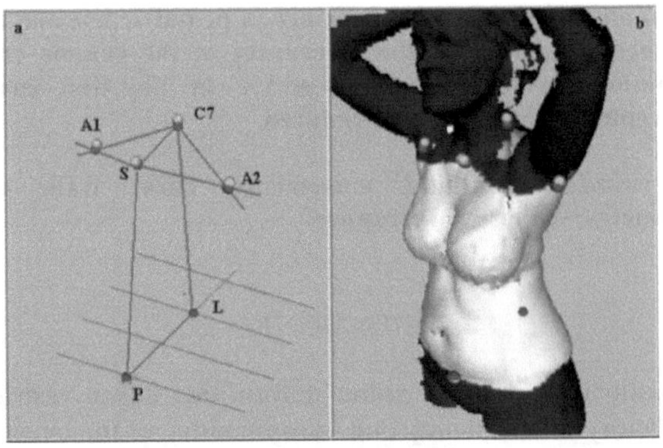

Fig. 3 Frame alignment scheme (a) and truncated volume (light grey) of interest (b)

The trunk was divided into left and right part by the surface P – L – C7 – S. This surface is close to but not equal to plane, since line segments P – S and L – C7 are not parallel. Deviation of this surface from plane is related to the rotation of the shoulders in relation to the pelvis due to the scoliosis deformity.

Calculations of trunk volumes were made by division of the whole trunk into 100 transversal slabs, perpendicular to the L – C7 axis. For each slab, the line, connecting line segments P – S and L – C7 was drawn to divide the slab into left and right parts. All 3D points, belonging to the slab, were projected into the slab's transversal plane and the approximating outline curve was drawn through projection points. The volume of the slab was calculated as numerically integrated surface of the slab multiplied by the slab thickness. Volumes of the left and right side slabs were calculated separately. The corresponding sums over all slabs give the estimations of the trunk left and right volume. The differences between volumes, calculated for inspiration and expiration frames were used as an estimate of the breathing volumes. As a result, the following parameters were obtained: 1) total trunk volume at inspiration, expressed in liters; 2) volumes of the left and right sides of the trunk, expressed in percents of the total trunk volume; 3) asymmetry index of the trunk, calculated as

$$\frac{right\ volume - left\ volume}{right\ volume + left\ volume}; \quad (1)$$

4) total breathing volume, expressed in liters; 5) left and right breathing volumes, expressed in percents of the total volume; 6) asymmetry index of the breathing pattern, calculated as

$$\frac{right\ breathing\ volume - left\ breathing\ volume}{right\ breathing\ volume + left\ breathing\ volume}. \quad (2)$$

To estimate the repeatability of the landmark selections and volume calculations, the whole procedure was repeated five times, using one particular patient as a model. The procedure was tested for the following modes: 1) manual selection of landmarks, identified by physically attached patient markers at the expiration frame and automated landmark alignments; 2) manual selection of landmarks, identified by physically attached patient markers both at expiration and inspiration frames; 3) marker free manual selection of landmarks at the expiration frame and automated landmark alignments; 4) marker free manual selection of landmarks both at expiration and inspiration frames.

To evaluate and visualize the movements of the chest wall, for each point of the expiration frame, the conjugate closest point at the inspiration frame was found. Afterwards, the projection of the vector, connecting two matching points on the direction of the normal to the expiration frame point was calculated and used as a measure of the chest wall breathing displacement; the onward movement was assumed to be positive. Obtained differences were visualized by color scale.

III. RESULTS AND DISCUSSON

Graphical representation of the breathing movements (fig.4) demonstrates ability of the method to reveal asymmetries in the chest and abdomen movements. At the presented image, more intensive coloring, observed at the patient right side, indicates more expressed movement of that side. Since the right side is the convex one for this patient, the obtained result seems to be in agreement with the known fact that the convex lung volume is larger than the concave one in scoliosis patients [15].

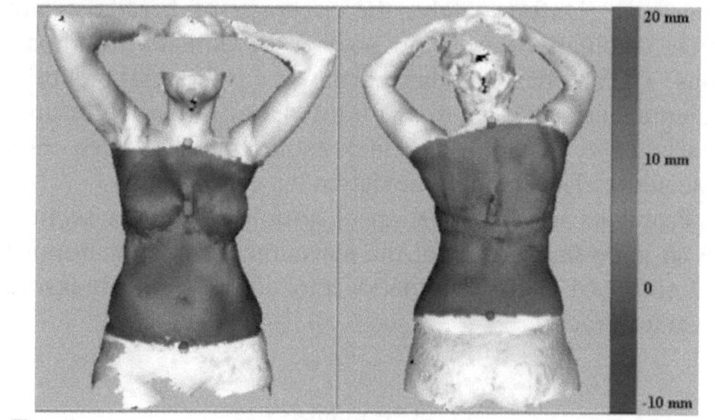

Fig. 4 Color map representation of the breathing movements during deep breathing.

Data on calculated volumes and measurement accuracy are summarized in Table 1, average value ± standard deviation (SD expressed in %).

Table 1 Calculated volume data

Parameter	Landmarks identification and alignment			
	With markers		Without markers	
	auto	manual	auto	manual
Total volume, l	19.4±0.2 (0.9%)	19.4±0.2 (1.1%)	19.8±0.4 (2.1%)	19.6±0.4 (2.1%)
Left volume, %	55.9±0.8 (1.4%)	55.5±0.5 (1.1%)	55.0±0.6 (1.0%)	54.7±0.8 (1.5%)
Right volume, %	44.1±0.8 (1.7%)	44.5±0.5 (1.2%)	45.0±0.6 (1.2%)	45.3±0.8 (1.8%)
Volumes asymmetry index, %	11.8±1.5 (12.9%)	11.0±1.1 (9.6%)	10.0±1.1 (11.3%)	9.4±1.7 (17.6%)
Total breathing volume, l	1.3±0.1 (5.8%)	1.4±0.5 (33.0%)	1.2±0.1 (7.0%)	1.2±0.3 (21.6%)
Left breathing volume, %	54.6±11.2 (20.6%)	57.2±5.8 (10.1%)	67.1±3.7 (5.5%)	57.9±8.6 (14.9%)
Right breathing volume, %	45.4±11.2 (24.7%)	42.8±5.8 (13.5%)	32.9±3.7 (11.1%)	42.1±8.6 (20.5%)
Breathing asymmetry index, %	9.1±22.5 (246.8%)	14.4±11.6 (80.7%)	34.2±7.3 (21.4%)	15.9±17.3 (108.6%)

The values of the total, left and right trunk volumes, calculated using different landmark selection and alignment methods are in a good agreement and may be calculated with a reasonable accuracy. The standard deviation was less than 2% in the worst case. The accuracy of volume estimation does not differ between automatic and manual alignment methods. The volume asymmetry index appeared to be very sensitive to the variation in landmark selection. The standard deviation was 9.6% in the best case.

In contrast, the accuracy of the breathing volume calculation decreases when landmarks are aligned manually. The standard deviations were significantly higher (P value of the Fisher test < 0.02) both for marker based, and marker free selections. Unfortunately, the accuracy of breathing volumes calculations was too poor for quantitative comparisons. The smallest standard deviation for the total breathing volume was 5.8%, whereas for the right / left volumes up to nearly 25%. Asymmetry index estimation was even more inaccurate. The standard deviation exceeded 100%.

Problems in identification of armpit landmarks seem to be an important factor for the inaccuracy. Better anatomical definition of these landmarks may improve the measurement accuracy.

IV. CONCLUSIONS

Proposed 3D optical scanner system allowsa dynamic acquisition of patient trunk 3D shapes without the use of markers. Chest and abdominal wall breathing movements can be quantified and visualized from the above 3D data.

The method allows estimation of the total trunk volume and its asymmetry with a reasonable accuracy. However, the evaluation of the breathing volume was not satisfactory probably due to problems in the landmark selection procedure. Further development is required to identify sources of inaccuracy and to improve the method.

ACKNOWLEDGMENT

The project was funded by the European Regional Development Fund (Project "Mobile Telemedicine Screening Complex", agreement Nr.2011/ 0007/ 2DP/ 2.1.1.1.0/ 10/ APIA/ VIAA/ 008), AOSpine Europe and Karolinska Institutet.

REFERENCES

1. Weinstein, S.L. (2001), The Pediatric Spine. Principles and practice, Lippincott William and Wilkins,
2. Koumbourlis, A. C. (2006). Scoliosis and the respiratory system. Paediatric Respiratory Reviews, 7(2), 152–160.
3. Kotani, T. . c, Minami, S. ., Takahashi, K. ., Isobe, K. ., Nakata, Y. ., Takaso, M, Moriya, H. . (2004). An Analysis of Chest Wall and Diaphragm Motions in Patients with Idiopathic Scoliosis Using Dynamic Breathing MRI. Spine, 29(3), 298–302.
4. Pehrsson, K. Danielsson, A. & Nachemson, A. (2001). Pulmonary function in adolescent idiopathic scoliosis: A 25 year follow up after surgery or start of brace treatment. Thorax, 56(5), 388–393.
5. Yuan, N. ., Fraire, J. A. ., Margetis, M. M. ., Skaggs, D. L. ., Tolo, V. T. ., & Keens, T. G. (2005). The effect of scoliosis surgery on lung function in the immediate postoperative period. *Spine*, *30*(19), 2182–2185.
6. Giordano, A., Fuso, L., Galli, M., Calcagni, M. L., Aulisa, L., Pagliari, G., & Pistelli, R. (1997). Evaluation of pulmonary ventilation and diaphragmatic movement in idiopathic scoliosis using radioaerosol ventilation scintigraphy. Nuclear Medicine Communications, 18(2), 105–111
7. Chun, E. M., Suh, S. W., Modi, H. N., Kang, E. Y., Hong, S. J., & Song, H.-R. . (2008). The change in ratio of convex and concave lung volume in adolescent idiopathic scoliosis: A 3D CT scan based cross sectional study of effect of severity of curve on convex and concave lung volumes in 99 cases. *European Spine Journal*, *17*(2), 224–229.
8. Kotani, T., Minami, S., Takahashi, K., Isobe, K., Nakata, Y., Takaso, M., Moriya, H. . (2002). Three-dimensional analysis of chest wall motion during breathing in healthy individuals and patients with seoliosis using an ultrasonography-based system. In *Studies in Health Technology and Informatics,* Vol. 91, pp. 135–139.
9. De Groote, Wantier, M., Cheron, G., Estenne, M., & Paiva, M. (1997). Chest wall motion during tidal breathing. *Journal of Applied Physiology (Bethesda, Md.: 1985)*, *83*(5), 1531–7.
10. Cala, S. J., Kenyon, C. M., Ferrigno, G., Carnevali, P., Aliverti, A., Pedotti, A., Rochester, D. F. (1996). Chest wall and lung volume estimation by optical reflectance motion analysis. *Journal of Applied Physiology*, *81*(6), 2680–2689.
11. Ricieri, D. V, & Rosário Filho, N. A. (2009). Effectiveness of a photogrammetric model for the analysis of thoracoabdominal respiratory mechanics in the assessment of isovolume maneuvers in children. *Jornal Brasileiro de Pneumologia: Publicaça Oficial Da Sociedade Brasileira de Pneumologia E Tisilogia*, *35*(2), 144–150
12. Yu, M. C., Liou, J. L., Kuo, S. W., Lee, M. S., & Hung, Y. P. (2012). Noncontact respiratory measurement of volume change using depth camera. *Conference Proceedings:. Annual International Conference of the IEEE Engineering in Medicine and Biology Society. IEEE Engineering in Medicine and Biology Society. Conference*, *2012*, 2371–2374.
13. Hamaoui, A., Gonneau, E., & Le Bozec, S. (2010). Respiratory disturbance to posture varies according to the respiratory mode. Neuroscience Letters, 475(3), 141–4.
14. P. J. Besl and N. D. McKay, "A method for registration of 3-D shapes," *IEEE Trans. Pattern Anal. Mach. Intell.*, vol. 14, no. 2, pp. 239–256, 1992.
15. E. M. . Chun, S. W. . Suh, H. N. . Modi, E. Y. . Kang, S. J. . Hong, and H.-R. . Song, "The change in ratio of convex and concave lung volume in adolescent idiopathic scoliosis: A 3D CT scan based cross sectional study of effect of severity of curve on convex and concave lung volumes in 99 cases," *Eur. Spine J.*, vol. 17, no. 2, pp. 224–229, 2008.

Address of the corresponding author:

Author: Alexei Katashev
Institute: BIN institute, Riga Technical University
Street: Kalku street 1
City: Riga
Country: Latvia
Email: katashev@latnet.lv

An Innovative Method for Standardizing Lean Management Approach in Hospitals

I. Lasorsa[1], G. Liuzzi[2], R. Calabrese[1], and A. Accardo[1]

[1] Dept. of Engineering and Architecture, University of Trieste, Trieste, Italy
[2] Ente Ecclesiastico Ospedale Generale Regionale "F. Miulli", Acquaviva delle Fonti, Bari, Italy

Abstract— **Lean management constitutes an emerging approach in the healthcare context, in order to increase quality of care and reduce costs. Nevertheless, literature shows the lack of a standardized method, especially in healthcare, due to the complexity of the involved processes.**

This paper proposes an innovative method for standardizing the lean management approach in Healthcare, in order to reduce wastes particularly in Hospitals. The integration of different techniques, such as IDEF0, and GQM, used in the process management is proposed, in order to let the overall method reproducible and repeatable. The application and feasibility of the method has been studied in an Italian Hospital.

Keywords — **Lean Management, Process Improvement, Waste reduction, Clinical Engineering, Healthcare.**

I. Introduction

Lean principles have been worldwide used in the service industries, in order to reduce wastes *(muda)*, improve processes, increase quality of services, and decrease costs. Lean management was introduced for the first time in the manufacturing sector (Toyota) [1], and it has been successful in waste reduction of the production process, delivering high-quality products. Its success has been proved in the automotive and electronic fields [2], too.

In the last decade, lean has been applied also in Healthcare services that significantly need of improvement in the domains of safety, effectiveness, timeliness, and appropriateness of care services [3], representing one of the most considerable trends in service industry [4]. Moreover, the improvement of care processes, which represents another direct consequence of the lean management implementation, produces reduction of costs and the consequent resources reallocation.

Several worldwide healthcare Organizations are currently using a lean approach. Nevertheless, a standardized method has not been yet defined, coherently with the lean management historical evolution (initial lack of written procedures for describing the method) [5].

Furthermore, the implementation of lean as a systematic approach in hospitals is not easy, due to the complexity of the care processes, the stakeholders involved [6], the specific needs and different cultures [7] of the Organizations. Moreover, one of the most challenging goals the healthcare systems have to deal with is the rapid growing and acquisition of new technologies [8]: people involved in the processes have to be continuously trained, and Standard Operating Procedures must be frequently updated. Another crucial aspect to be considered is the resistance of the operators to management changes, anytime a new process or technology is introduced.

For these reasons, hospital must be considered as a complex dynamic environment, and the introduction of a systematic approach must be studied and assessed carefully.

In order to add value to an Organization, it is essential to identify and eliminate waste, starting from the analysis of processes. The lean tool used for this purpose is the Value Stream Map (VSM), a method for analyzing the current state (Current Stream Map) and designing a future state for a process (Future Stream Map). Nevertheless, some of the metrics used in the VSM lack of a unique definition, especially in the healthcare context. This makes difficult to compare different studies in the literature, and to evaluate clearly the actual improvements of the Organization following the lean techniques application. Indeed, the goal of a metric should be obtaining objective, reproducible and quantifiable measurements, but this goal is not always reached, due to differences in the definitions of VSM metrics available in the literature.

In order to measure properties of software and its specifications, in the computer science context many software metrics (e.g., Halstead Complexity, DSQI, Robert Cecil Martin's software package metrics, Goal Question Metric) have been used. Particularly, Goal Question Metric (GQM) [9] is used to gather the measurement data and drive decision making and improvements, providing a support for the identification of the metrics starting from the definition of goals.

On the other hand, different functional modeling methods (e.g., function block diagram, HIPO and IPO, N2 Chart, IDEF0) are available in system engineering and they can be used as a support to the Value Stream Mapping. Particularly, IDEF0 is sometimes used as a support of lean management, since it allows mapping and analyzing complex interactions of a system [10].

In this paper, a standardized method, based on lean principles, supported by IDEF0 and GQM that can be used in any manufacturing and healthcare context is presented.

H. Mindedal and M. Persson (eds.), *16th Nordic-Baltic Conference on Biomedical Engineering*,
IFMBE Proceedings 48, DOI: 10.1007/978-3-319-12967-9_18

Moreover, the problem related to the lack of mathematical formula for defining some metrics used in the VSM is faced and a solution is proposed.

II. MATERIAL AND METHODS

A. Lean Thinking and Lean wastes

The main purpose of lean management is to create value for the end customer, through the elimination of wastes. Its philosophy derives mostly from the Toyota Production System (TPS), and it is focused on reduction of the original Toyota seven wastes (transport, inventory, motion, waiting, overproduction, over processing, defects) to improve overall customer value. Some other wastes have been added later on from other researchers, even though they have not been universally accepted [11].

The Value Stream Mapping is the lean management method for mapping the actions of the actors involved in a process. It employs standard symbols to represent items and processes, and key metrics are associated with each phase of the described process.

Moreover, different tools can be used and integrated at TPS operational level, such as Total Quality Management for assessing quality, Theory of Constraints for identifying constraints and restructure the rest of the organization around it, Six Sigma (6σ) and Statistical Process Control for assessing the variability of a process.

B. IDEF0

The IDEF0, a part of the IDEF family of modeling languages of software engineering, is a function modeling methodology that provides a representation of a process, combing texts, diagrams and graphic, detailing information about the input and output of an activity, how that activity is controlled, and the involved resources and mechanisms. It results in a hierarchical description that shows the links among the activities. The overall description of the process can be specified according to the level of the description needed by the Organization, or the complexity of the process itself.

C. GQM

The Goal Question Metric is a measurement system that can be divided into three levels:

- Goal (Conceptual level): it defines the main purposes of a work to be measured;
- Question (Operational level): it defines a set of questions useful for achieving the goals;
- Metric (Quantitative level): it defines a set of metrics for answering the questions in a measurable way.

The overall GQM diagram is shown in Figure 1.

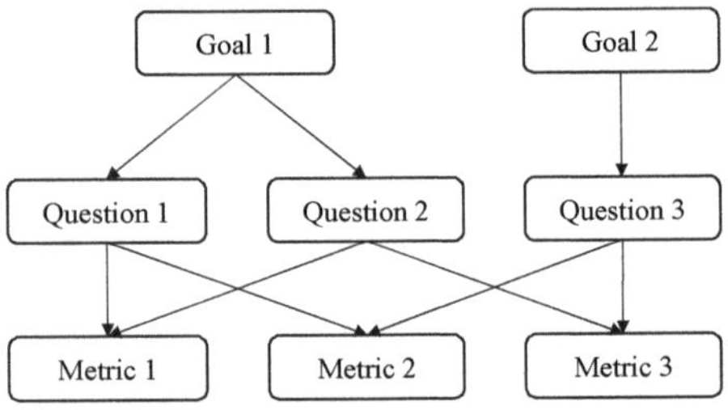

Figure 1 – GQM overall diagram

III. STANDARDIZING LEAN APPROACH

In order to improve an activity in accordance with the lean-thinking, an analysis of the overall process must be addressed. A team-leader must be identified for getting guidance and suggestions on directions. The team leader should have the capability to work in a multi-disciplinary team, for interacting with people with different backgrounds. She/he should also have an engineering background, for providing systematic and precise instructions to the team, in order to design and describe the process. Moreover, she/he should know and understand the healthcare environment as well as the most common related issues.

IDEF0 can be used to describe the process and identify the most problematic areas and the inter-relations among different activities (Figure 2). This phase must be addressed with the contribution provided by the actors involved in the process, who can describe the actual problems and, consequently, the real improvement areas. The team-leader must follow personally the process, in order to gain an external perspective of the main problems associated.

For identifying measurable metrics, the support of GQM can be valuable. Since the main goal of the lean-management is to reduce wastes, the reduction of *mudas* selected by the Organization, can be used as the Goals in the GQM. Questions must be found in order to reduce the wastes identified in a process, and measurable metrics (described through mathematical formula) must be defined.

As required by lean philosophy, the Current Stream Map can be designed by summarizing the most important phases of the process to be improved and selecting the metrics, identified through GQM and included in the *data box* (Figure 2). Through the analysis of the collected data, improvement solutions based on lean-techniques (e.g., Work Standardization, Poka-Yoke, Kanban, Spaghetti Diagram) can be implemented, and the Future Stream Map can be designed.

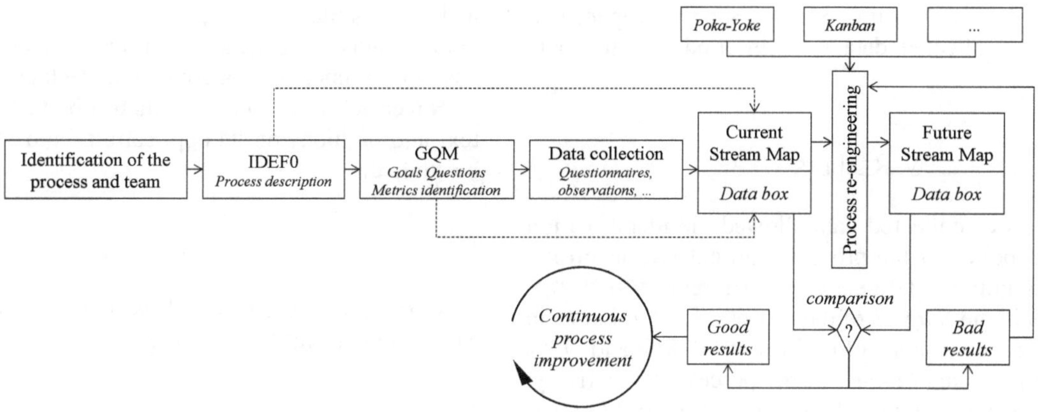

Figure 2 - Diagram block describing the proposed method for standardizing lean approach

Through the comparison between the metrics of the Current and the Future Stream Maps, the improvements of the process can be monitored during the time (Figure 2).

IV. CASE STUDY

In order to evaluate the described method, it was tested in the Operating Theater of the Italian Hospital F. Miulli. Particularly, the surgical tools process was studied. The criteria used for selecting this process among the others were related to the importance of the activity in a *core* process, the arising attention on the matter by the international patient safety organizations (e.g., JCI), and the complaints related to the surgical tools the top managers have been received from clinicians and nurses working in the Operating Theater. Preliminary, informal interviews were made with the actors involved in the process, and the professionals who wanted to give a contribution to improve the process were selected as part of the team.

The process was described with IDEF0, and detailed descriptions were made with respect to the most precarious activities (i.e. methodology for preparing the surgical tools containers, number and typology of surgical tools prepared before surgical operations, nonconformities).

The reduction of the seven *muda* (mentioned in the introduction), was set as goal in the GQM. Fifteen questions were identified (e.g. How many surgical tools prepared before the surgical operation are used during it? Is the inventory management appropriate? How much time is spent for non-added value activities? Are the storage areas located in appropriate places?). Seventeen metrics were defined (e.g., Cycle time, Uptime %, Percentage of used surgical tools, transport time). For each metric, a mathematical formula was provided. In the sake of example, some of the metrics were defined as follow.

Cycle time (*CT*):

$$CT = \frac{1}{N}\sum_{n=1}^{N}(t_{fn} - t_{in})$$

represents the average time for completing an activity of the process, being *N* the total number of observations, t_{in} and t_{fn} the initial and final time of the $n-th$ activity, respectively.

Non-Value Added time of the process (NVA):

$$NVA = \frac{S \cdot \bar{t}_s}{N}$$

related to one of the most problematic activity of the process (avoidable transports made by the operators looking for needed surgical tools not available in the Operating Rooms), defined as the spent time for doing avoidable tasks with non-added value; *S* is the total number of transports, and $\bar{t}_s$ is the average time of the transports.

Value-Added time (*VA*):

$$VA = CT - NVA$$

defined as the difference between the Cycle Time and the Non-Value Added time.

Uptime Percentage:

$$\%\,Uptime = \frac{VA}{CT} \cdot 100$$

defined as the percentage of the process that brings only added value.

Data related to twenty-three observations, one for each surgical operation, were collected.

The Current Stream Map was designed, through the identification of the most important activities of the process: preparation of the tools on the surgical cart in the Operating Room (*Stage 1*), utilization of the surgical tools (*Stage 2*), positioning of the instruments in the surgical case cart to be transported to the Central Sterile Supply Department (*Stage*

3), transport to the Central Sterile Supply Department (*Stage 4*). The collected data were inserted into the data boxes.

V. Results

The analysis of collected data allowed the identification of the critical points of the process. Particularly, according to the metrics initially defined, they were related to *% Uptime* = 71.2% in the *Stage 1* of the process; $\bar{t}_s$ =5 minutes in the overall process; nonconformities = 47.8% and 75% during the *Stage 1* and *Stage 3* of the process, respectively. The process has been reviewed, the Future Stream Map was designed, the new process was simulated and the economic saving was calculated (estimated to be 229.421 €), as it is discussed in the next session.

VI. Discussion

Through some improvement activities, such as the inventory of the surgical tools, and the new definition of the tools in the containers, is possible to increase the availability of the surgical tools, reduce the number of tools to be transported and to be processed by the Central Sterile Supply Department. From an economical perspective, an estimated hospital saving of 229.421 € could be produced by application of these improvements.

Moreover, the redefinition of the storage areas and the application of both 5S (a lean method for organizing a work space for efficiency and effectiveness [12]) and Visual Management (a lean visual control method for increasing the efficiency and effectiveness of a process by making the steps in that process more visible [13]) may also reduce the time spent for looking for the needed surgical tools (saving estimated to be 13.000 €).

Furthermore, the Cycle time would be reduced of the 33% during the preparation of the surgical tools, and the utilization of the tools would increase of 70% and 45% during the preparation of surgical case cart and the utilization phases, respectively.

Finally, nonconformities would be reduced of 47% and 75% during the preparation of the surgical tools cart and the positioning of the tools on the cart, respectively.

VII. Conclusions

The application of the method was a useful support for the study and for the application of the lean-management techniques. Structured metrics were provided and this makes possible the comparison between different studies. Good results were achieved in terms of process efficiency, cost, cycle time, and nonconformity reduction.

Nevertheless, the method should be tested in larger context and opinions on its applicability and efficiency should be studied.

Acknowledgment

Work partially supported by the University of Trieste, Master in Clinical Engineering.

References

1. J. P. Womack DT, Roos D, The Machine That Changed The World, New York: Rawson Associates, 1990
2. Voss CA, Alternative Paradigms for Manufacturing Strategy, International Journal Operations & Production Management, vol. 25, n. 12, 2005, pp 1211-1222
3. Houchens N, Kim CS, The Application of Lean in the Healthcare Sector: Theory and Practical Examples, Lean Thinking for Healthcare, Springer (2014)
4. Holm M, and Ahlstrom P, Lean Service: a literature review. In: Proceedings of the 17th International Annual EurOMA Conference, Portugal, June, 2010, pp 1-10
5. Brandão de Souza L, Pidd M, Exploring the barriers to lean health care implementation, Public Money & Management, 2011, pp 59-66
6. Kohn LT, Corrigan JM, Donaldson MS (2000) To err is human: building a safer health system. Washington, DC: Institute of Medicine, National Academies Press
7. Machado Guimarães M. Crespo de Carvalho J, Lean Healthcare across Cultures: State-Of-The-Art, American International Journal of Contemporary Research, vol. 2, n. 6, 2012, pp 187-206
8. Agha L, The effects of health information technology on the costs and quality of medical care, Journal of Health Economics, Vol. 34, 2014, pp 19–30
9. Basili VR, Caldiera G e Rombach HD, The Goal Question Metric Approach, in Encyclopedia of Software Engineering, Wiley, 1994
10. Muthiah KMN, Huang SH, A review of literature on manufacturing systems productivity measurement and improvement, International Journal of Industrial and Systems Engineering, vol 1, n. 4, 2006, pp 461-484
11. Mika G, Kaizen Event Implementation Manual, 5th edition
12. D. Sarkar, 5S for Service Organizations and Offices: A Lean Look at Improvements, ASQ, 2006
13. Greif M, The Visual Factory: Building Participation Through Shared Information, Productivity Press, 1991

Author: Irene Lasorsa
Institute: University of Trieste, Dept. of Engineering and Architecture
Street: Via Valerio, 10
City: Trieste
Country: Italy
Email: irene.lasorsa@phd.units.it

Harvesting Features of an Affordable Cognitive Gym for Seniors

C. Lassfolk[1], M. Linnavuo[1], S. Talvitie[2], M. Hietanen[3], and R. Sepponen[1]

[1] Health Factory/Department of Electrical Engineering and Automation, Aalto University, Espoo, Finland
[2] ORTON Foundation / ORTON Rehabilitation Ltd., Helsinki, Finland
[3] Department of Neurology, Helsinki University Central Hospital, Helsinki, Finland

Abstract—**First, this paper conducts a minireview of rehabilitation research in pursuance of concepts reusable in a cognitive gym. Second, it presents the discovered ideas and components as a framework for a preliminary cognitive gym for seniors. The cognitive gym offers an affordable, flexible, and safe training environment for seniors. This facility promotes systematic training of the concerted efforts of cognitive and motor skills.**

The longer life expectancy indicates an increase in the number of seniors. A cognitive gym constitutes a potential facility in senior residences, residential care buildings, and private homes. A multimodal training environment supports the seniors to preserve functionalities required for everyday life activities. This improves their quality of life and helps them remain autonomous.

Adaptable, individualized training programs consist of visual, auditory and motor elements. The exerciser encounters visual and auditory stimuli and distractors to which she responds by clicking a response button, by pointing at a certain object or by presenting a more complex motor pattern. The game-like character of the training motivates the person to exercise regularly and persistently.

Keywords—**cognitive training environment, rehabilitation, seniors**

I. Introduction

The demographics in many countries suggest an increase in the number of seniors and a longer life expectancy [1]. A long life does unfortunately not necessarily imply a good quality of life. In order to retain their autonomy and quality of life seniors need to maintain the functionality needed for everyday life activities.

During our research in cognitive stroke rehabilitation we realized that many of the rehabilitation tasks could also serve as keep-fit tasks for healthy seniors. Therefore, we harvested reusable concepts mainly in stroke rehabilitation research reports. After this brief literature review we listed the items that seemed most relevant for a low-budget training framework.

A. Literature review

Virtual reality: Virtual environments (VEs) or virtual realities (VRs) simulate real environments. Computers generate them in order to provide sensory input through one or more of the human senses. VRs in motor tasks are often characterized as three-dimensional, interactive, graphical simulations that monitor movement kinematics and provide haptic and force feedback. The experience of presence signifies that the user perceives the VE as realistic, and that the gap between imagination and reality diminishes. Such an immersive environment completely absorbs the person. Many immersive VEs utilize head-mounted displays (HMDs), which may, however, cause cybersickness. Liquid-crystal display (LCD) projectors and large wall screens constitute a relatively cheap and convenient setup which requires no glasses and still creates a sense of presence. [2-4]

Transfer: Current research commonly assumes that performance in an ecologically valid rehabilitation environment transfers more efficiently to the real world. Research hypothesizes that naturalistic VRs for rehabilitation which promote feedback, motivation, and repetition, stimulate neural plasticity in an optimal manner. For some motor learning tasks a VE may even outperform training in the real world. The gain of this transfer should inform future research and investment, but sufficient evidence has not yet been delivered. [3, 5-7]

Adaptability: In order to intensify the utilization of available information and communication technologies (ICT) as well as to increase rehabilitation intensity and decrease the workload of the therapists Caballero Hernández et al. (2011) have designed a set of cognitive tasks. These two-dimensional tasks adapt to the patient and her rehabilitation needs and provide objective performance data. The data analysis reveals whether the patient resides in the therapeutic range. The tasks adjust to the performance level of the patient, which thus directs the evolution of the therapy. [8] Besides the motivational factor this adaptability directly impacts learning. Interactive learning at the optimal individual pace appears to accelerate the overall process of learning. Healthy persons seem to learn a concrete skill faster, when they are allowed to interactively adapt the

H. Mindedal and M. Persson (eds.), *16th Nordic-Baltic Conference on Biomedical Engineering*,
IFMBE Proceedings 48, DOI: 10.1007/978-3-319-12967-9_19

video demonstration of the task to their information acquisition needs. [9] Alankus et al. [10] investigate the demand for adjustability, customizability, and configurability required in rehabilitation games, which stimulate the motivation of the exerciser. They identify factors that influence the level of challenge experienced by the rehabilitee and try to build games that adapt to users at different levels of recovery.

Serious games: The vast supply of videogames has spawned the idea to use games not only as entertainment. So called serious games develop, inter alia, in the areas of education, government, military, and health. They serve in cognitive as well as motor tasks. They strive to inherit the engaging, challenging, and rewarding characteristics from the entertainment games while merging these with a central training objective. PlayMancer [11] represents a game development platform project, which strives to reduce the game development cycle and cost and thus to increase the availability of augmented realities and three-dimensional games for serious edutainment (education/entertainment) purposes. This system integrates traditional input channels such as joystick, keyboard, and mouse with novel ones such as speech, touch, biosensors, and motion tracking. [11-14]

Everyday life activities: Many research projects examine the rehabilitation of everyday life activities in a safe and synthetic computer-generated environment. Christiansen et al. [15] investigate the feasibility of an immersive, virtual kitchen in which patients prepare a meal. Naveh et al. [16] examine the impact of a non-immersive VR system on a street-crossing task. Persons with an attentional impairment are requested to cross the street in a safe and vigilant manner. Rand et al. [17] develops a video-capture VR system which implements a therapeutic virtual mall environment for stroke patients. Shopping in the mall serves as cognitive and motor training. Cameirão et al. [5] study the efficacy and transfer of motor rehabilitation in a rehabilitation gaming system based on VR technology. A virtual classroom serves as a tool for assessing attention impairments in children [18]. Costa et al. develop a virtual city concept for reacquisition of cognitive abilities [6]. Dores et al. [7] define another virtual city for rehabilitation of acquired brain injury victims. In the simulated city the participant plans a set of given daily tasks and then implements the plan by moving in the city on foot or by car, taxi, subway, train or bus in a cost and time efficient manner. This intervention presumably improves executive functions, visuospatial processing, attention, and memory. [7] Kizony et al. [19] explore the performance of and relationship between cognitive and sensorimotor skills in VEs. The project experiments with virtual ball games and soccer which embed a real-time video of the user into the game.

Home-based rehabilitation: In order to improve cost-efficiency health care organizations look for home-based rehabilitation solutions or so called telehealth or telerehabilitation. This implies that the patient resides at home and receives services from a remote therapist. The therapist can define and tune the tasks in the training program, coach the patient and monitor the rehabilitation progress. This saves travelling time and enables the therapist to attend to more patients. [20, 21] Pareto et al. [21] explore the feasibility of solutions for the two major challenges in home-based rehabilitation of stroke patients: the low motivation of the patients to perform repetitive and intense training and the need for frequent follow-up meetings. Deutsch et al. [20] investigate the technical and patient performance in an integrated VR – telerehabilitation system. Harley et al. [22] develop an affordable video-gaming based motor-rehabilitation task suited for exercise at home or in the clinic. Costa et al. [6] identify the possibilities offered by virtual realities and the World Wide Web to disseminate therapeutic environments at low cost even to rural regions.

Satisfaction: Rehabilitation systems based on VR technology constitute potential occupational spaces, where older stroke survivors can sense competence, creativity, and satisfaction with life, besides experiencing pleasure and flow. [23]

Interactive tabletop: Augstein et al. [24] explore the feasibility of using an interactive tabletop as a means of man-machine communication in neurorehabilitation. This mixed-reality environment recognizes finger, hand, and physical object multi-touches. Therefore, the proposed environment enables individuals as well as small groups of people to perform mutual and social tasks on the table.

II. Materials and Methods

From the literature review presented in section I.A we extracted features that were frequently mentioned or otherwise seemed central to a generic cognitive gym.

III. Results

This section presents characteristic features of a cognitive training environment grouped into three categories: General, equipment, and tasks.

A. General

Several modalities: The exercises generate multiple sensations and reactions. Such different modalities encompass distinct auditory and visual stimuli as well as oral and motor responses. If possible, these sensations and reactions mimic ecologically valid real-world situations. A VR embeds the exerciser into the task. This fosters learning and motivation. [2, 13]

Safety: Security considerations prevail over all other properties. The exercisers need the opportunity to challenge their skills in a risk-free environment. Practicing car driving or safe street crossing serves as a good example of risky real-world tasks. [16]

Availability and accessibility: The site needs to serve when people want to practice. Furthermore, wheelchair and walker users need access to the site and the equipment.

Cost: The system has to offer the service at an affordable price. Both private and public entities struggle with restricted budgets. A rehabilitation system presumably increases the quality of life for the individual and saves money for the public health services. However, it appears unlikely that the full amount of the savings will return into the preventive health maintenance. Thus, a rehabilitation system available for the majority of the population even in the developed countries has to consist of reasonably-priced components. [6]

Telerehabilitation: A remote personal trainer can supervise the training session over a videoconferencing connection. [3] However, the senior must master the equipment in the gym on her own.

B. Equipment

Computer: A computer controls the training system. For a low-cost training environment a basic desktop or laptop computer suffices. More sophisticated graphics and simulations may need specialized hardware.

Display: A training environment requires one or more displays. Many of the tasks will operate with a normal LCD or a video projector, whilst some more advanced exercises might utilize HMDs. [3]

Sounds: Some tasks will probably require wireless headphones or loudspeakers. Sounds can add realism to a simulated environment or they can serve as stimuli and distractors in tasks.

Mouse, keyboard: An ordinary keyboard and a mouse control the computer. Some exercises exploit additional input devices such as gaming mice, joysticks, response buttons, or data gloves.

Gesture detection: A video camera detects gestures performed by the exerciser. The Xbox Kinect [25] exemplifies such a device with a camera that preferably resides in front of the gesturing person.

Balance board: Seniors can benefit from balance training. A Wii balance board [26] or a similar inexpensive device suits the need for monitoring balance exercises. [3]

C. Tasks

To further independence and quality of life for seniors a cognitive gym offers a variety of multimodal, controllable tasks consisting of serious games. Such games train and preserve everyday life activities. The exercises improve a smooth cooperation of cognitive and motor skills. The tasks focus on a pedagogical purpose, but they inherit the engaging properties and the sense of presence from the gaming world. Seniors experience the exercises as pleasant and rewarding. [8, 22]

In some tasks the user sits, in some she stands and in other she moves around. Motions, gestures, button presses and touches serve as response methods. The performance requirement of the exercise changes with the achievements of the person. Thus, a too difficult task decreases the performance requirement and a too easy task raises it. Different users perform different training programs. The programs change over time, as the skills develop, as needs change, or as a means to sustain motivation. The training platform and the facility support a multitude of tasks requiring a variable amount of space and additional equipment. [3]

The personal trainer can define and control the training program remotely. Anyhow he does not attend all training sessions. He can follow the progress of the training remotely, change the exercises and adjust the degree of difficulty. Nonetheless, the system can also automatically adjust the requirement level. The exerciser receives feedback about her performance directly from the system. The feedback and the reward strongly influence satisfaction and motivation. This drives further sustained repetition of the exercises. [10]

IV. Discussion

We conducted a brief literature review of rehabilitation environments used in research in order to create a picture of features relevant to a cognitive training environment suited for seniors. We decided to refer to this concept of voluntary training in terms borrowed from the physical exercise context in order to assign a positive note to it. Thus, we call the facility a cognitive gym and the therapist a personal trainer.

V. Conclusions

A safe, affordable and versatile cognitive training environment provides a good starting point for private or shared domiciliary use. Further research may define in more detail practical implementations of such an environment.

Acknowledgments

The Finnish Cultural Foundation and the Finnish Funding Agency for Technology and Innovation (Tekes) have supported this work.

References

1. Lunenfeld B (2008) An aging world-demographics and challenges. Gynecological Endocrinology 24:1-3 DOI 10.1080/09513590701718364
2. P. Gamito, J. Oliveira, D. Morais, et al., NeuAR–A review of the VR/AR applications in the neuroscience domain, Augmented Reality - some Emerging Application Areas, InTech, pp. 131-154 DOI 10.5772/26135
3. Holden M K (2005) Virtual environments for motor rehabilitation: review. Cyberpsychology & Behavior 8:187-211 DOI 10.1089/cpb.2005.8.187
4. Weiss P L, Rand D, Katz N, et al. (2004) Video capture virtual reality as a flexible and effective rehabilitation tool. J. Neuroeng Rehabil. 1:12 DOI 10.1186/1743-0003-1-12
5. Cameirão M S, Badia S B, Oller E D, et al. (2010) Neurorehabilitation using the virtual reality based Rehabilitation Gaming System: methodology, design, psychometrics, usability and validation. Journal of Neuroengineering and Rehabilitation 7:48 DOI 10.1186/1743-0003-7-48
6. da Costa R M E M, de Carvalho L A V, de Aragon D F (2000) Virtual city for cognitive rehabilitation, Proceedings of the 3rd International Conference on Disability Virtual Reality and Associated Technologies. Alghero, Sardinia, 2000, pp 299-304
7. Dores A R, Miranda M J, Carvalho I P, et al. (2012) Virtual city: Neurocognitive rehabilitation of acquired brain injury, Information Systems and Technologies (CISTI), 2012 7th Iberian Conference On, 2012, pp 1-4
8. Caballero Hernandez R, Martinez Moreno J M, García Molina A, et al. (2011) 2D-Tasks for Cognitive Rehabilitation.
9. Schwan S, Riempp R (2004) The cognitive benefits of interactive videos: learning to tie nautical knots. Learning and Instruction 14:293-305 DOI :10.1016/j.learninstruc.2004.06.005
10. Alankus G, Lazar A, May M, et al. (2010) Towards customizable games for stroke rehabilitation, Proceedings of the SIGCHI Conference on Human Factors in Computing Systems, 2010, pp 2113-2122 DOI 10.1145/1753326.1753649
11. Conconi A, Ganchev T, Kocsis O, et al. (2008) Playmancer: A serious gaming 3d environment, Automated Solutions for Cross Media Content and Multi-Channel Distribution, 2008. AXMEDIS'08. International Conference On, 2008, pp 111-117 DOI 10.1109/AXMEDIS.2008.29
12. Ma M, Bechkoum K (2008) Serious games for movement therapy after stroke, Systems, Man and Cybernetics, 2008. SMC 2008. IEEE International Conference On, 2008, pp 1872-1877 DOI 10.1109/ICSMC.2008.4811562
13. Omelina L, Jansen B, Bonnechere B, et al. (2012) Serious games for physical rehabilitation: Designing highly configurable and adaptable games, Proc. 9th Intl Conf. Disability, Virtual Reality & Associated Technologies, 2012,
14. Rego P, Moreira P M, Reis L P (2010) Serious games for rehabilitation: A survey and a classification towards a taxonomy, Information Systems and Technologies (CISTI), 2010 5th Iberian Conference On, 2010, pp 1-6
15. Christiansen C, Abreu B, Ottenbacher K, et al. (1998) Task performance in virtual environments used for cognitive rehabilitation after traumatic brain injury. Arch. Phys. Med. Rehabil. 79:888-892
16. Naveh Y, Katz N, Weiss P (2000) The effect of interactive virtual environment training on independent safe street crossing of right CVA patients with unilateral spatial neglect, Proceedings of the Third ICDVRAT, 2000, pp 243-248
17. Rand D, Katz N, Kizony R, et al. (2005) The Virtual Mall: A Functional Virtual Environment for Stroke Rehabilitation. Annual Review of Cybertherapy and Telemedicine 3:193-198
18. Rizzo A A, Buckwalter J G, Bowerly T, et al. (2000) The virtual classroom: a virtual reality environment for the assessment and rehabilitation of attention deficits. CyberPsychology & Behavior 3:483-499 DOI 10.1089/10949310050078940
19. Kizony R, Katz N, Weiss P L (2004) Virtual reality based intervention in rehabilitation: Relationship between motor and cognitive abilities and performance within virtual environments for patients with stroke, Proc. 5th Int. Conference on Disability, Virtual Reality and Associated Technologies, 2004, pp 19-26
20. Deutsch J E, Lewis J A, Burdea G (2007) Technical and patient performance using a virtual reality-integrated telerehabilitation system: preliminary finding. Neural Systems and Rehabilitation Engineering, IEEE Transactions On 15:30-35 DOI 10.1109/TNSRE.2007.891384
21. Pareto L, Johansson B, Ljungberg C, et al. (2011) Telehealth with 3D games for stroke rehabilitation. International Journal on Disability and Human Development 10:373-377 DOI 10.1515/IJDHD.2011.062
22. Harley L, Robertson S, Gandy M, et al. (2011) The design of an interactive stroke rehabilitation gaming system. Human-Computer Interaction.Users and Applications 167-173 DOI 10.1007/978-3-642-21619-0_22
23. Reid D (2003) Virtual reality: An enabling environment for occupational performance of leisure activities for the older stroke survivor. University of Toronto Medical Journal 80:192-194
24. Augstein M, Neumayr T, Ruckser-Scherb R, et al. (2013) The fun. tast. tisch. project: Interactive tabletops in neuro-rehabilitation, Proceedings of the 2013 ACM International Conference on Interactive Tabletops and Surfaces, 2013, pp 313-316 DOI 10.1145/2512349.2514591
25. Kinect for Windows features at http://www.microsoft.com/en-us/kinectforwindows/discover/features.aspx (accessed 2014-05-28)
26. What is Wii fit plus? at http://wiifit.com/what-is-wii-fit-plus/ (accessed 2004-05-28)

Address of the corresponding author:

Author: Christina Lassfolk
Institute: Aalto University
Street: Otakaari 7B
City: Espoo
Country: Finland
Email: christina.lassfolk(at)aalto.fi

MammoAid

Nanying Liang, Srinath Sridharan, and James Mah Tzia Liang

Institute for Infocomm Research, Agency for Science, Technology and Research, Singapore
liangny@i2r.a-star.edu.sg

Keywords— **mammographic density, segmentation; artificial intelligence, breast cancer, enter up to five keywords and separate them by commas.**

I. BACKGROUND

Breast cancer is the most commonly diagnosed cancer in females worldwide. At present, mammography is the best non-intrusive way for breast cancer detection. It is well established that mammographic density (MD) is an independent and robust factor for breast cancer risk assessment [1]. Breast cancer risk assessment is useful in personalizing mammography screening and in helping patients make informed decision about the ways of reducing breast cancer risk. Currently, MD estimation is done manually or semi-automatically (Cumulus [2], for example). This leads to intra- and inter-reader variations in MD estimation, and hence limits its application in breast cancer risk assessment.

II. METHODS

In pursuit of objective MD estimation and subsequently applying MD risk factor for breast cancer risk assessment, we made an effort to standardize MD estimation by developing a fully automatic MD analysis procedure and later implemented it in the software MammoAid. The key features of the software MammoAid can be summarized as follows: i) The software is compatible with analogue and digital mammograms. ii) The software is fully automatic for breast area and pectoral muscle segmentation. iii) The software provides a means for objective MD estimation. iv) Artificial intelligence technique is embedded in the software for learning the personal shading preference of a radiologist by updating the threshold for dense tissue segmentation. v) The software is able to process multiple mammograms in an efficient high-throughput manner. Here we brief the software, followed by the short descriptions of its techniques and methods used.

The software provides a user-friendly GUI to analyse mammograms and to generate reports. Behind the GUI, we implemented an automatic procedure for MD estimation. We further introduced the artificial intelligent techniques for emulating radiologist's expert opinion of dense tissue segmentation. The procedure for MD estimation is summarized in Fig. 1, in which a series of steps are involved including contrast enhancement, noise suppression, breast area extraction, label removal, pectoral muscle removal and dense tissue segmentation. These steps can be explained as follows:

- Mammograms are greyscale X-ray images with intensity values ranging from 0-255. In order to enhance image representation, we linearly stretched the histogram of the original mammogram to fully cover the range from 0-255. For analogue mammograms, we further applied a series of top-hat transforms to suppress the noise due to digitization.
- Breast mask was extracted using an improved version of image extended minima [3] by adopting an adaptive threshold, which was estimated from the histogram of the intensity for each mammogram. The label consisting of information about view and laterality was further removed from the breast mask.
- Mammograms consist of two typical views: craniocaudal (CC) and mediolateral oblique (MLO) view. For the MLO mammograms, pectoral muscle is observed and removed using the more recent and efficient algorithm proposed by Kwok et al [4]. The key features of Kwok's method are: i) iterative and precise identification of the region of pectoral muscle, which takes into account the varying size of pectoral muscle for different studying subjects. ii) automatic identification of the pectoral muscle edge through detecting the interruptive intensity changes in the intensity profiles.
- After extracting breast area from the background including pectoral muscle being removed, now we can carry out dense tissue segmentation using a threshold method, the Niblack's algorithm [5]. Based on the mean and standard deviation of the intensity of breast area, the algorithm automatically estimated the optimal threshold for dense tissue segmentation. The MD was then estimated as the ratio of the number of pixels in the dense tissue to the number of pixels in breast tissue.

H. Mindedal and M. Persson (eds.), *16th Nordic-Baltic Conference on Biomedical Engineering*,
IFMBE Proceedings 48, DOI: 10.1007/978-3-319-12967-9_20

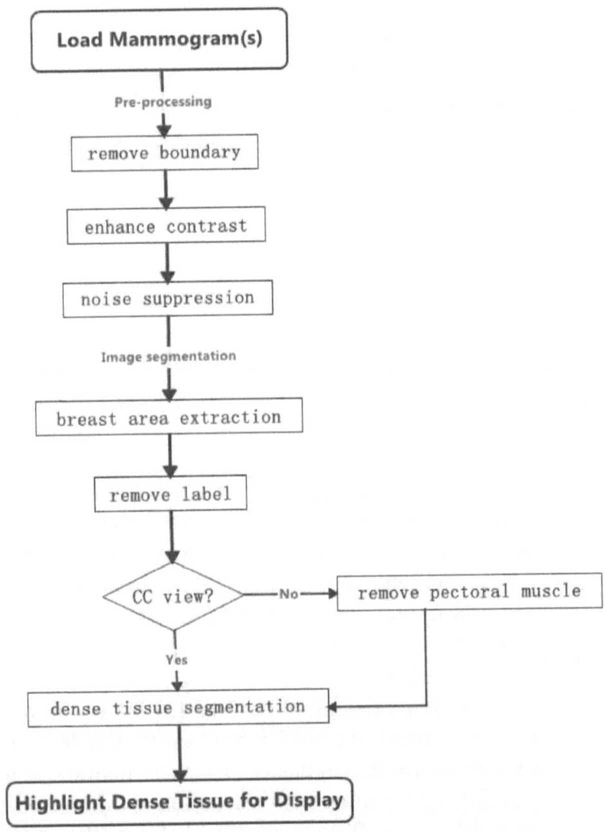

Fig. 1 Flowchart for the implementation of the software MammoAid.

We noticed that, in the Niblack's algorithm, there was a sensitivity parameter, which controls the amount of tissue to be segmented as dense tissue. Based on this observation, we further proposed a second-order polynomial regression model for modelling this parameter. In this way, the software can automatically learn and adapt the shading preference of an individual radiologist, and ultimately minimize the intra- and inter-reader variations in MD estimation.

III. RESULTS

For the purpose of performance evaluation, we used a set of 1438 digitized mammograms. Each mammogram has the dimension of 600 pixels in width and 833 pixels in height, and a horizontal and vertical resolution of 72 dpi with an image depth of 8. For each mammogram, it comes with a suggested MD ratio from a doctor by drawing on the mammogram and calculating the ratio manually. These suggested ratios were used as reference in evaluation. The simulation results showed that our software, when compared with doctor's reference, achieved an accuracy of 87% and Pearson's correlation coefficient of 0.95 in MD ratio estimation.

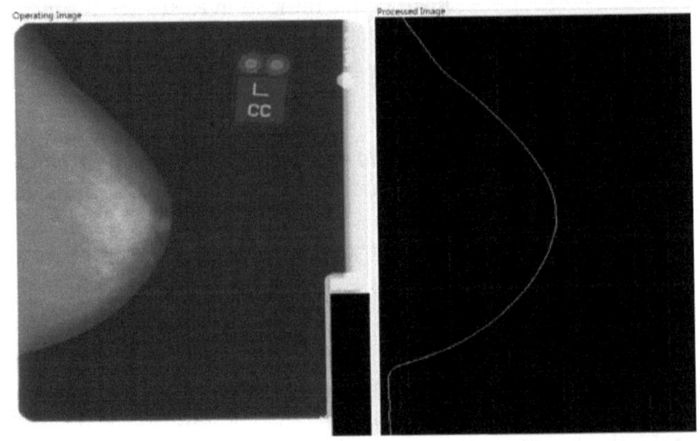

Fig. 2 For demonstration purpose of the interaction and adaptation feature of the software, in the initial point, we artificially forced the software to shade none of dense tissue for this CC mammogram by setting the sensitivity parameter as 1.70, leading the MD ratio estimation to zero.

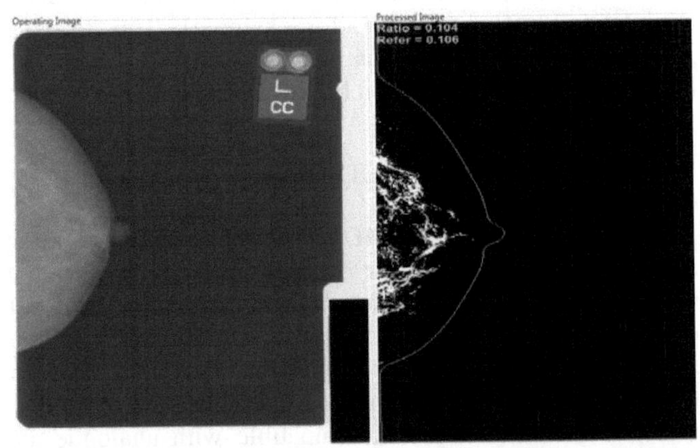

Fig. 3 After training with 16 mammograms, the software learned the shading behavior of a doctor. The trained model was then applied to a new CC mammogram and resulted in a MD ratio estimation of 0.104, which was close to the doctor's reference of 0.106.

The software can also learn the shading preference of a radiologist through the regression model of the sensitivity parameter aforementioned. When the training starts, it is assumed that the software doesn't know how to segment the dense tissue at all. W artificially tuned the sensitivity parameter such that the MD ratio was initially incorrectly estimated by the software as shown in Fig. 2. From this initial point, we trained the software to learn the shading preference of a radiologist from a training dataset of mammograms. The simulation showed that the regression model was able to correctly learn the sensitivity parameter in an iterative manner. As illustrated in Fig. 3, the software segmented the dense tissue accurately after training with 16 mammograms.

IV. Conclusions

In summary, our software can accurately shade breast tissue and pectoral muscle as well as objectively estimate MD ratio in a fully automatic manner. On top of this automatic implementation, we added the feature of emulating radiologist's expert opinion of mammograms through the inclusion of artificial intelligence. It paves the way for the MD estimation standardization through a large-scale clinic trial in the future.

Acknowledgment

The authors thank Prof Chia Kee Seng and his team from National University of Singapore (NUS) for granting permission to use their mammogram dataset.

References

[1] C.M. Vachon, C.H. van Gils, T.A. Sellers, K. Ghosh, S. Pruthi, K.R. Brandt and V.S. Pankratz, "Mammographic density, breast cancer risk and risk prediction," Breast Cancer Researchc, vol. 9, no. 6, pp. 217-225, 2007.

[2] J.W. Byng, N.F. Boyd, E. Fishell, R.A. Jong and M.J. Yaffe, "The quantitative analysis of mammographic densities," Physics in Medicine and Biology, vol. 39, no. 10, pp. 1629-1638, 1994.

[3] P. Soille, Morphological Image Analysis: Principles And Applications, Springer-Verlag, 1999.

[4] S.M. Kwok, R. Chandrasekhar, Y. Attikiouzel and M.T. Rickard, "Automatic pectoral muscle segmentation on mediolateral oblique view mammograms," IEEE Trans. Medical Imaging, vol. 23, no. 9, pp. 1129-1140, September 2004.

[5] W. Niblack, An Introduction To Digital Image Processing. Prentice Hall, 1986

Parallel Measurements of in-vivo Skin Autofluorescence Lifetimes and Photobleaching Rates

A. Lihachev[1], I. Ferulova[1], J. Spigulis[1], and D. Chorvat[2]

[1] Institute of Atomic Physics and Spectroscopy/ Biophotonics Laboratory, University of Latvia, Riga, Latvia
[2] International Laser Center/ Bratislava, Slovakia

Abstract— **Experimental methodology for parallel measurements of *in-vivo* skin autofluorescence (AF) lifetimes and photo-bleaching dynamic has been developed and tested. The AF lifetime decay distributions were periodically collected from fixed tissue area with subsequent detection of the fluorescence intensity decrease dynamic at different time shifts after the pulse excitation. Temporal distributions of skin AF lifetimes and bleaching dynamic were collected and analyzed by means of commercial time-correlated single photon counting system. Details of the equipment and data processing are described as well as some measurement results that confirm the feasibility of the proposed technology.**

Keywords— ***in-vivo* skin, autofluorescence, lifetime, photobleaching.**

I. INTRODUCTION

Laser induced time-resolved autofluorescence (AF) spectroscopy represents a promising adjunctive technique for in-vivo tissue diagnostics. Biological and biomedical applications of the autofluorescence spectroscopy have been successfully applied in cell biology and clinical diagnostics for detection of abnormal tissues [1-5]. The most important endogenous fluorophores are molecules widely distributed in cells and tissues, like proteins containing aromatic amino-acids, NAD(P)H, flavins and lipo-pigments [6,7]. The radiative lifetime of each fluorophore is unique, so fluorescence lifetime measurements can provide specific information on fluorphore content and distribution in the tissue. Tissue fluorescence lifetime value depends on environmental factors e.g. experimental setup, fluorophore localization, skin pH balance, viscosity, temperature etc. [8-10]. Selective analysis and separation of individual fluorophores underlying multi-exponential decays of *in-vivo* skin autofluorescence still is challenging problem.

In addition, tissue autoflurescence usually shows the photo-bleaching effect which is caused by chemical modification of the fluorochrome due to repeated excitation/emission states [11]. Due to mechanism of photobleaching is unclear, the purposeful use of photo-bleaching in biomedical applications are used very rarely. However, some of the authors successfully applied bleaching effect in Raman spectroscopy for improving the Raman spectra quality by decreasing of tissue fluorescence background [12, 13]. Temporal decrease of skin autofluorescence (AF) intensity can be well described by a double exponential function (1). Under long-term continuous/pulse excitation, the main decrease of the intensity I occur within the first 10–15 seconds, followed by relatively slow decrease which strives to a constant intensity level A:

$$I(t) = a_1 exp^{-t/T_1} + a_2 exp^{-t/T_2} + A \quad (1);$$

T_1 represents the fast phase of AFPB, T_2, the slow phase; a_1, a_2, and A are constants, and t is time [13,14]. The tissue AF photo-bleaching mechanism is still under discussion; however, one can assume that each tissue fluorophore might have its own specific AF bleaching rate, and any changes in the surface fluorophore composition would induce corresponding changes in the AF decay distribution and bleaching rates.

II. MATERIALS AND METHODS

The main focus of this study is to assess the possibility and potential of a combined time-resolved fluorescence spectroscopy method including parallel analysis of tissue autofluorescence lifetimes and photo-bleaching rates. For this purpose an experimental setup of parallel point measurements of skin AF lifetimes and photo-bleaching rates was assembled (Fig.1). The setup comprises a pico-second excitation laser (PicoQuant: pulse half-width 59 ps, 405nm, mean power density ~2mW/cm^2, mod. LDH-D-C-405) laser controller, monochromator, photon counting detector with temporal resolution of 180ps (mod. Becker&Hickl, PMC-100-4), data processing system "time-correlated single photon counting" with time resolution 6.6 ps, (TCSPC, mod. Becker&Hickl SPC-150), and a fiber optic probe. AF lifetimes of position stabilized healthy skin were measured during 3 minutes.

H. Mindedal and M. Persson (eds.), *16th Nordic-Baltic Conference on Biomedical Engineering*,
IFMBE Proceedings 48, DOI: 10.1007/978-3-319-12967-9_21

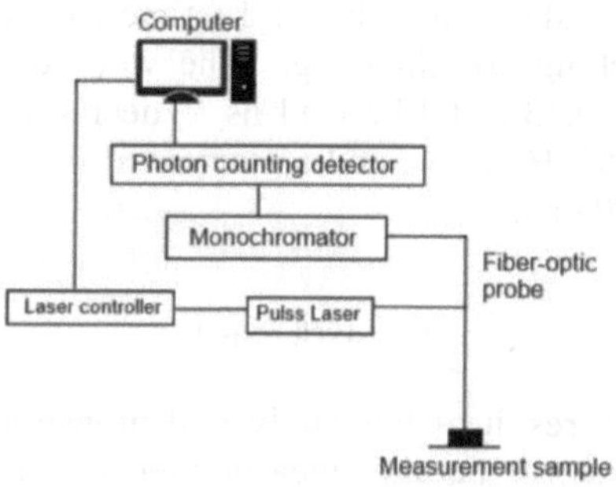

Fig. 1 Experimental setup for parallel point measurements of skin autofluorescence lifetimes and photo-bleaching dynamic.

Skin AF lifetimes were collected every 10 seconds. Totally during the 3 minute cycle 18 measurements (Fig.2) of tissue AF decay distributions were registered. Each measurement contained temporal distribution of AF decay and a number of registered photons collected during the 10 seconds. Due to the AF photo-bleaching process, the number of collected photons for each subsequent measurement decreases, thereby giving the opportunity to construct the AF photo-bleaching curves. Autofluorescence lifetimes were collected at 480 nm wavelength using Becker&Hickl software SPCM. For the analysis of AF lifetimes obtained, the multi-exponential fluorescence decay distribution model was applied:

$$f(t) = \sum_{i=0}^{n} a_i exp^{-t/\tau_i} + c \quad (2),$$

were f(t) is AF intensity at time moment t after the excitation pulse, n is number of decaying species in the exponential sum, and c is a background level of light in each particular case. The applied model comprises the lifetimes of the exponential components, τ_i, and the amplitudes of the exponential components, a_i.

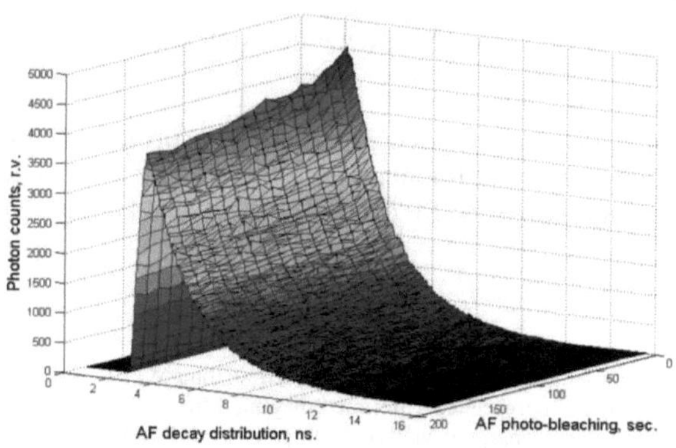

Fig. 2 *In-vivo* skin autofluorescence lifetime decay distributions during 3 minutes of pulse 405 nm excitation with mean power density 2mW/cm^2.

The program uses a deconvolution technique in conjunction with the measured instrumental response to function to obtain corrected autofluorescence decay curve in respect to instrument response. Thus, the set-up allows measuring the lifetimes up to 0.5 ns. The measurements were taken from 3 different spots of healthy skin at the inner part of forearm. For further analysis, the obtained AF decay distributions were approximated applying three-exponential decay model (n=3 (2)). Totally 5 volunteers with different skin photo types were involved into the study.

III. RESULTS

Table 1 represents the measured averaged values of healthy in-vivo skin AF lifetimes and their relative amplitudes. In all cases the healthy skin AF lifetimes and their relative distributions for all volunteers were nearly equal. The standard deviations (SD) show relatively low deviations from the average values.

Table 1. The averaged autofluorescence lifetime components τ_i, and, their relative amplitudes a_i of healthy skin for different volunteers

	a_1	τ_1	a_2	τ_2	a_3	τ_3
Average	69,26	1,52	22,12	6,02	8,6	8,10
Standard deviation (SD)	3,40	0,13	1,33	1,03	2,47	1,68

Each measurement can be also characterized by decreased number of collected photons during the 3 minute measurement cycle. Figure 3 demonstrates such decrease at different time shifts from the exiting pulse. As shown, the dynamics of collected AF photons in the ranges from 1… 2 ns, 5…6 ns and 8…9 ns are very similar and the mean photo-bleaching rate can be determined.

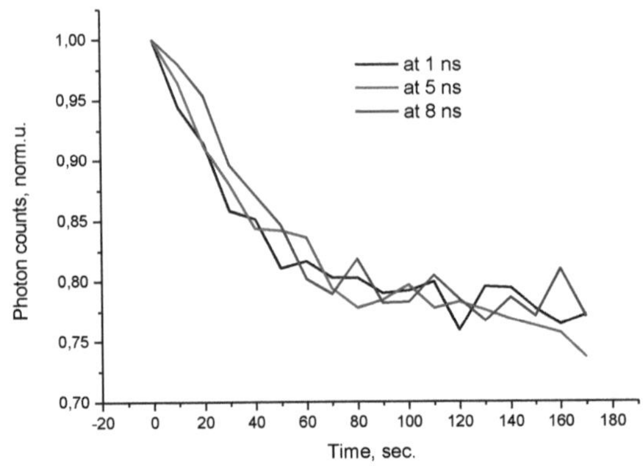

Fig. 3 Autofluorescence intensity decrease at different time shifts after pulse excitation with mean power 2 mw/cm^2.

However, in the case of ten time's stronger (20 mw/cm^2) excitation, the bleaching process becomes more intensive, and less homogeneous. In particular, during the 6 minutes of excitation the number of collected photons in the time gate 1…2 ns decreased approximately by 60%, in the time gate 5…6 ns by~ 55%, and in the time gate 8…9 ns - by~45% (Fig.4).

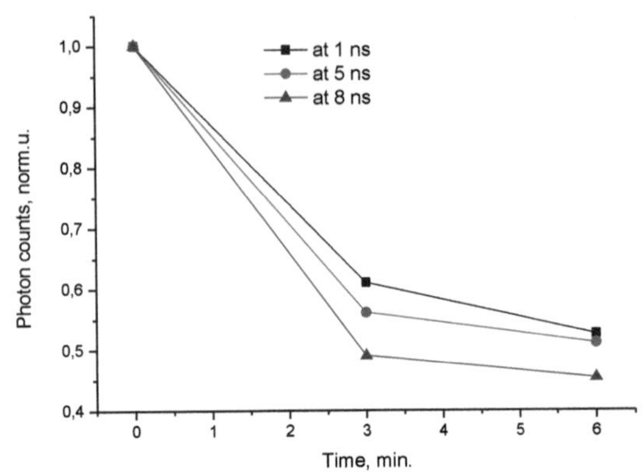

Fig. 4 Autofluorescence intensity decrease at different time shifts after pulse excitation with mean power 20 mw/cm^2.

This result shows that the AF lifetime components during photo-bleaching are changing. The value of τ_1 decreased from 1.52 ± 0.13 to 1.13 ± 0.11 ns, τ_2 decreased from 6.02 ± 1.0 ns to 4.95 ±0.62 ns, and τ_3 decreased from 8.1 ±1.68 ns to 5.97 ± 1.03 ns.

IV. DISCUSSION

The main result of this study is demonstration of possibility to take for parallel measurements of tissue fluorescence lifetimes and photo-bleaching rates by means of a commercial system. Healthy in-vivo skin can be better characterized by two dynamic parameters -autofluorescence lifetime and photo-bleaching rate. Furthermore, photo-bleaching efficiency can be estimated at different time shifts after the pulsed excitation. The observed good agreement between the photo-bleaching curves related to different time shifts at the mean excitation power density of 2mW/cm^2 (Fig.3) may be associated with dominating contribution in the resulting distribution by one specific fluorophore. However, under higher excitation power densities (20 mW/cm^2) the bleaching process becomes inhomogeneous, showing different photo-bleaching rates at different time shifts – it apparently indicates to some skin fluorophore content changes during the photo-bleaching process.

V. CONCLUSIONS

The proposed method demonstrates good perspectives for selective analysis and separation of individual tissue fluorophores underlying FLT and bleaching analysis. The proposed method further can be developed from point measurements to the imaging mode allowing for parallel visualization of tissue fluorescence lifetimes and bleaching rates.

ACKNOWLEDGMENT

This work was funded by the FP-7 projects “LaserLab Europe” (grant agreement n° 284464) and “Fotonika-LV” (grant agreement n° FP7-REGPOT-CT-2011-285912).

REFERENCES

1. Leng-Chun C, William R. Lloyd III, Ching-Wei C, Dhruv S et al. (2013) Fluorescence Lifetime Imaging Microscopy for Quantitative Biological Imaging. Method Cell Biol 114:457-488.

2. Tadrous PJ, Siegel J, French PMW et al. (2006) Time-resolved optical imaging provides a molecular snapshot of altered metabolic function in living human cancer cell models. Opt Express 14:4412-4426.
3. Rinaldi AG, Desplancq FD, Sibler A-P et al. (2013) The use of fluorescent intrabodies to detect endogenous gankyrin in living cancer cells. Exp Cell Res 319, 6:838–849.
4. Stamp GWH. (2003) Fluorescence lifetime imaging of unstained tissues: early results in human breast cancer. J Pathol 199:309-317.
5. Galletly NP, McGinty J, Dunsby C et al. (2008) Fluorescence lifetime imaging distinguishes basal cell carcinoma from surrounding uninvolved skin. Brit J Dermatol 159:152-161.
6. Roberts M.S, Dancik Y, Prow T.W et al. (2011). Non-invasive imaging of skin physiology and percutaneous penetration using fluorescence spectral and lifetime imaging with multiphoton and confocal microscopy. Eur J Pharm Biopharm 77, 3:469-488.
7. Blackwell J, Katika KM, Pilon L et al. (2008) In vivo time-resolved autofluorescence measurements to test for glycation of human skin. J Biomed Opt 13(1):014004 DOI: 10.1117/1.2830658.
8. Chorvat D, Chorvatova A. (2009) Multi-wavelength fluorescence lifetime spectroscopy: A new approach to the study of endogenous fluorescence in living cells and tissues. Laser Phys Lett 6, 3: 178-179.
9. Islam M. S, Honma M, Nakabayashi T et al. (2013) pH Dependence of the Fluorescence Lifetime of FAD in Solution and in Cells. INT J Mol Sci 14:1952–63.
10. Dmitrovsky E, Mycek M. A, Pitts J. Fluorescence lifetime spectrometer (fls) and methods of detecting diseased tissues. World Patent: WO2002069784 A2. Issued date September 12, 2002.
11. Lippincott-Schwartz J, Altan-Bonnet N, Patterson GH. (2003) Photobleaching and photoactivation: following protein dynamics in living cells. Nat Cell Biol 5:S7–S14.
12. Darvin ME, Meinke MC, Sterry W, Lademann J. (2013) Optical methods for noninvasive determination of carotenoids in human and animal skin. J Biomed Opt 61230 DOI: 10.1117/1.JBO.18.6.061230.
13. Wanga H, Zhaoa J, Leea A M.D et al. (2012) Improving skin Raman spectral quality by fluorescence photobleaching. Photodiagn Photodyn 9,4:299-302.
14. Zeng H, MacAulay C, McLean DI, Palcic B et al. (1998) The dynamics of laser-induced changes in human skin autofluorescence--experimental measurements and theoretical modeling. Photochem Photobiol 68(2):227-36.

Author: Alexey Lihachev
Institute: Institute of Atomic physics and Spectroscopy, Biophotonics laboratory, University of Latvia
Street: Raina blvd.19
City: Riga
Country: Latvia
Email: lihachov@inbox.lv

A 4-AFC Study Comparing Ultrasound Machines Using a Greyscale Phantom

R. Lorentsson[1,2], N. Hosseini[1], J.-O. Johansson[1], W. Rosenberg[1], B. Stenborg[1], L.G. Månsson[1,2], and M. Båth[1,2]

[1] Department of Medical Physics and Biomedical Engineering, Sahlgrenska University Hospital, Gothenburg, Sweden
[2] Department of Radiation Physics, Institute of Clinical Sciences, Sahlgrenska Academy, University of Gothenburg, Sweden

Abstract— **Two ultrasound machines, one high-end and one ordinary, were compared in terms of their ability to reproduce low-contrast objects using the same probe. Images containing 4 mm objects of four different contrasts were collected from a greyscale phantom at 35-42 mm depth. Six observers participated in a 4-alternative forced choice study based on 120 images. At this certain depth and object size the proportion of correct responses was higher (statistically significant) for the high-end machine at three of four contrast levels, indicating the possibility to discriminate between ultrasound machines using a limited number of images of a greyscale phantom. However, the number of images and number of observers needed are larger than usually used for constancy control.**

Keywords— **Ultrasound, phantom observation, detection, 4-AFC.**

I. Introduction

Ultrasound is a common medical diagnostic technique in modern healthcare. The ultrasound machine park in a hospital consists of a wide range of ultrasound systems. There are all from low end ultrasound systems to expensive high end with several probes and specially designed software for certain examinations. This, together with the fact that some of the machines are in the end of their life cycle and some in the beginning typically results in a large variety of ultrasound equipment in a hospital. The brightness mode (B-mode) scanning produces an image of the tissue where each echo is presented as a bright dot, the stronger echo, the brighter dot. It is not obvious how the quality of the B-mode image can be quantified in an objective and clinically relevant way. The reason for wanting to quantify the image quality is both to see if a machine degrades over time and be able to compare different machines. The B-mode image has three quantitative objectives [1]

- To visualize anatomy on as fine scale as possible
- To detect and, if possible identify, the presence of limited tissue size masses (lesions) within or adjacent to a reference tissue
- To detect, observe or measure movement of a structure in time.

The present paper focuses on the second of the three, i.e. the possibility to detect low contrast objects of different contrast. The purpose of the study was to compare two different ultrasound machines from the same manufacturer, one high end and one ordinary, in terms of their ability to reproduce low contrast and taking into account human observer variability.

II. Method

A. Machines and settings

The two ultrasound machines evaluated were a high-end GE Logiq9 (GE Healthcare, Milwaukee, USA) purchased 2006 (L9), and the more ordinary GE Logiq P5 (LP5), purchased 2007. In order to isolate the capability of the machine itself, the same probe was used for both machines (curved array C4 probe, (2-5 MHz)). The Abdomen-setting was used for the machines and only the gain, depth and focus were adjusted for best viewable image. The L9 used 4 MHz and the LP5 5 MHz for the Abdomen-setting. A single focus was set and positioned for the actual object depth in each image.

H. Mindedal and M. Persson (eds.), *16th Nordic-Baltic Conference on Biomedical Engineering*, IFMBE Proceedings 48, DOI: 10.1007/978-3-319-12967-9_22

B. Phantom and Image Acquisition

To evaluate objects of different contrast, the greyscale phantom CIRS 047 (Computerized Imaging Reference Systems, Incorporated, Norfolk, USA), was used. The phantom contains three different object sizes, 2.4, 4 and 6.4 mm, formed as cylinders. Every size has seven different contrast levels, anechoic, -9 dB, -6 dB, -3 dB, + 3 dB, +6 dB and +9 dB. The depth varies for the different sizes: 1-6 cm for 2.4 mm, 2- 9 cm for 4 mm and 3-12 cm for 6.4 mm. In the present study, only the 4 mm objects were used. The phantom was placed on a router table with ability to move the phantom in two directions with a precision of 0.1 mm. The probe was held by a clamp and in the same position while the phantom could move freely under the probe. Water was used as coupling medium. To get independent images of the objects according to the surrounding speckle the phantom was moved both sideways and along the long side. For each signal the phantom was placed in three different locations sideways 15 mm from each other and ten steps of 2 mm along the long side (three different angles for each slice position). In this way 30 images of the 4 mm objects for each of the contrast levels -6, -3, +3 and +6 dB were collected at 35-42 mm depth. The images were saved as DICOM images and transferred to a personal computer.

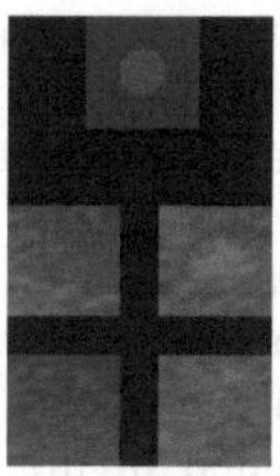

Fig. 1 An example of the 4AFC test object. The object on top indicates to the observer the size, position and contrast relative to the background.

C. Observer study

Using MATLAB R2013b (The Mathworks, Inc., Massachusetts, USA), the acquired images were used to produce a data set suitable for a 4-alternative forced choice (4-AFC) study. Around each signal, three background regions were extracted from the image. The four regions (1 signal region+3 background regions) were used to produce 4-AFC images, where the position of the signal image was randomized. As an aid, a reference square containing an object of the same size and contrast as the signal was shown on top of each image (Fig. 1). Six observers participated in the 4- 4-AFC study, five medical engineers and one medical physicist. The instruction to the observers was to determine which of the four squares that contained the signal. No time limitation was set, and the observers were free to alter zoom and window/level. ViewDEX [2, 3] was used for presenting the images in random order for each observer and recording the observers' choices. The ambient lightning was kept at a low constant level, and the images were presented on a DICOM-calibrated screen, EIZO Radiforce RX 320 (EIZO Corporation, Ishikawa, Japan). In total, each observer analyzed 120 images.

D. Statistical analysis

The proportion of correct responses (P) and the difference of P, P_{dif} between the two machines was calculated.

Bootstrap simulation was used to estimate the uncertainty. P and P_{dif} was bootstrapped 10 000 times for random combinations of observers and images for each contrast level. The percentiles 2.5 and 97.5 from the simulated data were used as 95% confidence interval (CI).

III. Result

Table 1 The average performance (proportion of correct responses, P) for the 6 observers in the 4-AFC study.

Contrast	L9 P (95% CI for P)	LP5 P (95% CI for P)
-6 dB	0.97 (0.92 - 1.00)	0.89 (0.80 - 0.96)
-3 dB	0.59 (0.47 - 0.71)	0.55 (0.43 - 0.68)
3dB	0.83 (0.68 - 0.94)	0.66 (0.51 - 0.78)
6dB	0.98 (0.92 - 1.0)	0.90 (0.78 - 0.98)

The result of the 4-AFC study for the two machines are presented in Table 1. At all contrast levels, the L9 had a higher value of P although the uncertainty was relatively large compared to the differences. Nevertheless, the difference was statistically significant at all contrast levels except – 3 dB (Fig 2).

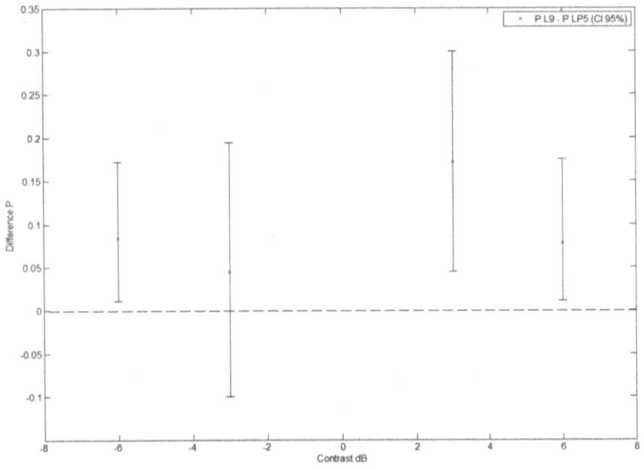

Fig. 2 The difference in P between L9 and LP5. The error bars represent 95% CI.

IV. Discussion

In the present study the ability of ultrasound machines to reproduce low-contrast object has been investigated. A 4-AFC study was conducted in which human observers evaluate images of a grayscale phantom. A statistically significant difference between the two included machines was found for three of four contrast levels, indicating a higher performance for the high-end machine in this task. The clinical relevance of this result depends on e.g. the validity of the detection task, how the phantom is constructed and the study design. These issues will be discussed below.

A. Images

The images in a multiple alternative forced choice should be statistically independent [4]. The method used here, produced 3 different angles from the object to the center of the transducer when the phantom was moved 15 mm in the lateral direction, and 10 different slices of the objects when the phantom was moved 2 mm in the elevation direction. Based on visual inspection, it was judged that the speckle changed enough when moving the phantom in the described directions to treat the images as independent, but no other analysis of the independence has been performed.

Regarding the location of the object in the signal image, this should be exactly known by the observer in an AFC study. In the present study there were some difficulties in establishing the true location of the signal in certain images, hence an uncertainty was introduced for the observers. However, the possible misalignment was small enough for this problem to be deemed of limited value.

B. Phantom properties

To use objects formed as cylinders to evaluate the ability to detect spherical lesions have been questioned [5, 6], since the effect of the slice thickness on the image is not taken in account. To avoid this problem the best way to perform this kind of comparison with real observers would probably be to collect a number of speckle-independent images with spherical low contrast objects from a phantom containing spherical lesions of different sizes, contrasts and depths. To our knowledge, none of the phantoms from the large manufacturers has a phantom like this in their product catalog. Either the spheres are just in one contrast (anechoic), or else it contains just one sphere of each contrast and size at most 3 depths, which makes it difficult to collect enough independent samples.

Another difference between using spherical objects and cylindrical is that phantoms containing spherical objects are more sensitive to the position of the probe in order to produce accurate images. The center of the beam has to align with the plane of the spheres in two directions to align with the center of the spheres. The position is important for the cylindrical objects as well but an error in the position affects what is being reproduced much more with the spheres. The presented method is a compromise between collecting many independent images of the same objects and the bias of the partial volume effect when using cylinders instead of spheres, when using a commercial available ultrasound phantom.

Regarding the clinical relevance of lesion size, a retrospective clinical study of liver lesions [7] has shown that the findings start between 4 and 5 mm. For this reason, the 2.4-mm signals were excluded in the present study. Furthermore, the visibility of the 6.4-mm signal was too high for relevant inclusion in a 4-AFC study. The 4-mm lesions included both had suitable visibility and a clinically relevant size. A study investigating the characteristics of intra-abdominal cystic masses showed that a variation in the number of internal echoes could be found in all the cystic masses examined [8]. It is therefore an advantage to include a range of contrast levels when studying the ability of a machine to reproduce low-contrast details.

The properties of different tissue mimicking materials in ultrasound phantoms have been studied [9]. The acoustic velocity in Zerdine™ (used in CIRS 047), remained constant (+- 3 m/s) but the attenuation were found to be nonlinear with increasing frequency, which could affect penetration depth at higher frequencies. As long as the compared objects are far from the penetration depth this effect would probably be minimal. The backscatter properties of the tissue mimicking material in the CIRS 047 is not mentioned in the specification from the manufacturer other than "scatter controlled independently from attenuation". The backscatter coefficient is difficult to measure accurately and therefore is rarely reported in the literature.[10] Nevertheless, the amount of scatter in the phantom results in images with a soft tissue-like appearance. The absence of other anatomical signs that the examiner take into account when searching for lesions makes a phantom study less valid, but an advantage is that it is easier to perform since the observers need no medical background, and the images are easy to make in large amounts. Also, the ability of machine to reproduce low contrast objects is an important property.

C. Uncertainty

The precision of an observer in a 4-AFC test depends on the number of observations and the probability of a correct answer. The relative standard deviation (s (d′)/d') for one observer, P = 0.8 and 30 images is 0.18 [4]. By dividing by the RMS of number of observers [11] the relative standard error for present study is $\frac{0.18}{\sqrt{6}} = 0.073$. This simple estimation of the uncertainty of the results agrees reasonably well with the size of the confidence interval determined using bootstrapping, although the latter method more accurately takes into account all existent sources of variability.

V. Conclusion

The results of the present study indicate that even if inter-observer variability is correctly taken into account it is possible to discriminate between ultrasound machines, in terms of their ability to reproduce low-contrast details, using a limited number of images of a greyscale phantom. However, the number of images and number of observers needed are larger than usually used for constancy control.

References

1. Hill C R, Bamber J C, Crawford D C et al. (1991) What might echography learn from image science? Ultrasound Med Biol 17:559-575
2. Borjesson S, Hakansson M, Bath M et al. (2005) A software tool for increased efficiency in observer performance studies in radiology Radiat Prot Dosimetry 114:45-52
3. Hakansson M, Svensson S, Zachrisson S et al. (2010) VIEWDEX: an efficient and easy-to-use software for observer performance studies Radiat Prot Dosimetry 139:42-51
4. Burgess A E (1995) Comparison of receiver operating characteristic and forced choice observer performance measurement methods Medical Physics 22:643-655
5. Kofler J M, Jr., Lindstrom M J, Kelcz F et al. (2005) Association of automated and human observer lesion detecting ability using phantoms Ultrasound Med Biol 31:351-359
6. Madsen E L (1991) Ultrasound focal lesion detectability phantoms Medical Physics 18:1171
7. Hall T J (1995) Ultrasound contrast-detail analysis: A comparison of low-contrast detectability among scanhead designs Medical Physics 22:1117
8. Hill M, and Sanders R C (1978) Gray scale B scan characteristics of intra-abdominal cystic masses Journal of Clinical Ultrasound 6:217-222
9. Browne J E, Ramnarine K V, Watson A J et al. (2003) Assessment of the acoustic properties of common tissue-mimicking test phantoms Ultrasound in Medicine & Biology 29:1053-1060
10. Culjat M O, Goldenberg D, Tewari P et al. (2010) A review of tissue substitutes for ultrasound imaging Ultrasound Med Biol 36:861-873
11. Tapiovaara M J, and Sandborg M (2004) How should low-contrast detail detectability be measured in fluoroscopy? Medical Physics 31:2564

Author: Robert Lorentsson
Institute: Sahlgrenska University Hospital
Street: Vita stråket 12
City: Gothenburg
Country: Sweden
Email: robert.lorentsson@vgregion.se

Mechanistic Modelling Investigates the Neural Basis behind the Hemodynamic Response in fMRI

K. Lundengård[1,2], G. Cedersund[3,4], F. Elinder[3], and M. Engström[1,2]

[1] Division of Radiological Sciences, Department of Medicine and Health Sciences, Linköping University, Linköping, Sweden
[2] Center for Medical Image Science and Visualization (CMIV), Linköping University, Linköping, Sweden
[3] Department of Clinical and Experimental Medicine, Linköping University, Linköping, Sweden
[4] Department of Biomedical Engineering, Linköping University, Linköping, Sweden

Abstract— **This work serves as a basis for a new type of fMRI analysis, which is based on a mechanistic interpretation of the hemodynamic response to synaptic activity. Activation was measured in the visual cortex of 12 healthy controls and ordinary differential equation models were fitted to the time series of the hemodynamic response. This allowed us to reject or refine previously proposed mechanistic hypotheses. This is the first attempt to describe the hemodynamic response quantitatively based on recent neurobiological findings. This mechanistic approach stands in contrast to the standard phenomenological description using the gamma variate function.**

Keywords— **functional magnetic resonance imaging (fMRI), blood oxygen level dependent (BOLD) response, mechanistic modeling, ordinary differential equations (ODE), neurovascular coupling.**

I. INTRODUCTION

This work serves as a basis for a new type of fMRI analysis, which is based on a mechanistic interpretation of the hemodynamic response to synaptic activity. Functional MRI measures the changes in oxygenation of different regions of the brain in response to neural activity, called the hemodynamic response. The classical way of analyzing such data is based on a phenomenological description of the hemodynamic response, using the gamma variate function. Such phenomenological descriptions fail to provide and make use of the full biological understanding of the system. In contrast, we here present a mechanistic model, which is based on previously proposed feedback signals involved in the hemodynamic response.

II. MATERIALS AND METHODS

12 healthy subject, mean age = 25.25 years (std = 4.73), performed a visual stimulation task where they were shown words of different colors during fMRI in a Philips Ingenia 3T MR scanner (TR=2s, TE=35ms, FA=75° resolution = 3x3x3 mm^3). Images from each subject were preprocessed (re-align, co-register, and smooth) using SPM8 [1]. The images were analyzed using the conventional correlation model in order to identify the peak activation in the visual cortex. The time series of the hemodynamic response were extracted from the un-normalized images of each individual in a 3x3x3 voxel cube around the peak activation in the visual cortex. The time series were then averaged over the group. The mathematical models were implemented as ordinary differential equations using the MATLAB SystemsBiology ToolBox. The model parameters were obtained from data by minimizing the chi square function.

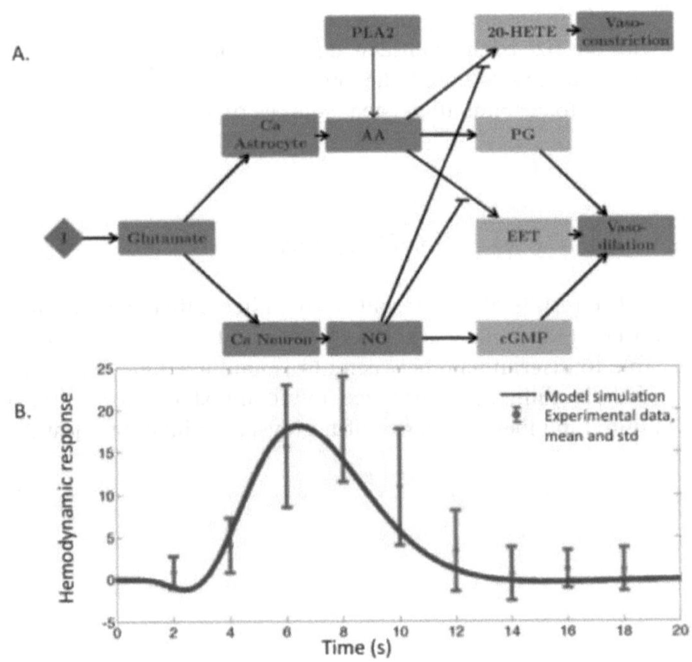

Fig. 1 A. The neurobiological model, constructed from ref [1]. I = input / stimulation, Ca = calcium ions, AA = arachidonic acid, PLA2 = phospholipase A2, 20-HETE = 20-hydroxy-eicosatetraenoic acid, PG = prostaglandins, EET = epoxyeicosa-trienoic acids. B. Model fit to hemodynamic response in the visual cortex of 12 healthy controls.

H. Mindedal and M. Persson (eds.), *16th Nordic-Baltic Conference on Biomedical Engineering*, IFMBE Proceedings 48, DOI: 10.1007/978-3-319-12967-9_23

III. Results

Two hypotheses were tested: (1) a metabolic model describing the blood flow in response to increased metabolic activity, and (2) a neurobiological model describing the neurovascular coupling [2] (fig 1A). Both models can describe some of the qualities commonly found in the literature, such as the initial dip [3] and the delayed rising phase and the fall of the hemodynamic response. The neurobiological model (2) can also describe the post-peak undershoot.

IV. Conclusions

As a proof-of-concept, we have shown that a mechanistic model based on theories of the neurovascular coupling has the ability to describe the hemodynamic response. The best fit to data is currently given by the neurobiological model, but future work includes further optimization and testing of the metabolic model. This is the first attempt to obtain a mechanistic and quantitative description of the hemodynamic response. This open the door to a new type of analysis of fMRI data that in the future can be used for studying brain connectivity and its relation to brain disorders.

However, to get improved estimates of the values of these mechanistic parameters, we will make use of mixed effect modeling, especially concerning their region-region and subject-subject variations. This will increase our understanding of what type of neural activity is actually measured by what we call "activity" in fMRI.

Acknowledgment

The research council of southeast Sweden (FORSS) and Linköping University are acknowledged for financial support.

References

1. Penny W, Friston K, Ashburner J, Kiebel S, Nichols T (2006) Statistical Parametric Mapping: The Analysis of Functional Brain Images, Academic Press
2. Attwell D, Buchan AM, Charpak S, Lauritzen M, Macvicar BA, Newman EA. (2010) Glial and neuronal control of brain blood flow. Nature. 11;468(7321):232-243
3. Hu, Yacoub. (2012) The story of the initial dip in fMRI. NeuroImage 62;1103-1108

Author: Karin Lundengård
Institute: Department of Medical and Health Sciences
Street: Linköpings universitet, 581 83
City: Linköping
Country: Sweden
Email: karin.lundengard@liu.se

Visualization of Real Time MR Guided Hyperthermia

David F. Marshall

The Research School of Computer Science, The Australian National University

Abstract— **I predict that improved methods of visualizing magnetic resonance guided (MRg) hyperthermia will be needed within four to eleven years. This will happen as a result of: increasing refresh rates for MR thermometry; improving understanding of thermal damage mechanisms; the limits of human perceptual abilities; and, the limits of current visualization methods. I establish a case for prioritizing the research of new interfaces. If new interactive visualizations are not developed in-time, advancing MR and medical technologies will out-pace the ability of software systems to fully utilize those advances.**

Keywords— **MR guided, hyperthermia, visualization**

I. INTRODUCTION

A. Overview

Magnetic Resonance Imaging (MRI) has been increasingly used in order to guide and assess the progress of surgical therapies over at least the last 20 years. Hassenbusch et al[1] was an early example. One particular category of such applications of MR guidance has been the use of MR thermometry to guide local hyperthermia, as with Quesson et al[2].

Hyperthermia includes treatments such as: Laser Interstitial Thermal Therapy (LITT); High Intensity Focussed Ultrasound (HIFU); and, Radio Frequency Ablation (RFA). Improved accuracy and timeliness of decision-making during hyperthermia have motiving health benefits, including: more thorough treatment of tumors; reduced unintended damage to surrounding tissue; and, reduced length of procedures.

The essential problem, in using real-time thermometry to guide local hyperthermia (such as LITT), is to halt treatment as close as possible to, but not beyond, the desired effect. Treating too far might result in unacceptable damage to healthy or sensitive anatomy, while incomplete treatment may lead to the need for re-treatment or reduced positive outcomes.

MR-guided Laser Interstitial Thermal Therapy (MRg LITT) is one example of local hyperthermia. The two primary such systems in use at this time[3] are the Visualase Thermal Therapy System[4], and the Monteris Medical NeuroBlate® (formerly AutoLITT®) System.

B. Thermal Damage Mechanisms

The various mechanisms by which excess heat can damage living tissue are complex. Some are direct mechanisms, such as the coagulation of blood or the carbonisation of tissue. Others are indirect, such as an increased sensitivity to cytotoxic drugs. The primary mechanism of cell death near the centre of the treatment volume (or centre of heat application), in the case of LITT, will be coagulation. At the periphery of the volume, however, tissue death will occur due to the accumulated thermal dose.

In his discussion of standard ways to calculate hyperthermia dosage, Atkinson[5] predicted that numerical values for cytotoxicity will improve as clinical experience increases. This anticipated improvement of our understanding of thermal damage mechanisms is one of the factors that will drive the need for better interfaces, in addition to improved thermal measurement. Despite the amount of research effort that has been applied to increasing our understanding of these processes, Milleron and Bratton[6] found that the mechanisms of heat shock-induced cell death remain poorly understood. As our understanding of this interaction is refined, we might be motivated to apply heat with smaller tolerances for error.

C. Real-Time Visualization of Therapeutic Lesion Formation

LITT involves the laser heating of tissue, such as in a brain tumor, to the point of causing cell death by mechanisms such as coagulation. This area of dead tissue is known as a therapeutic lesion. MR thermometry has been shown to be extremely useful in monitoring the resulting tissue temperatures during LITT. LITT is a well established, and evolving, treatment methodology. As early as 1998, Skinner et al[7] identified that interstitial heating had become the most commonly used method of heating tumors. This approach has since continued to be refined. Tyc and Wilson[8] said that NeuroBlate® allows surgeons to selectively direct the thermal treatment, according to the specific nature of the tissue in the target area.

Since the lesion and tumor volumes can be irregularly shaped, and since the lesion volume's growth is not entirely predictable, monitoring the convergence of the surfaces of these two volumes is a non-trivial visualization task.

H. Mindedal and M. Persson (eds.), *16th Nordic-Baltic Conference on Biomedical Engineering*,
IFMBE Proceedings 48, DOI: 10.1007/978-3-319-12967-9_24

Since the thermal characteristics of tissue change after coagulation, it is desirable to start and stop heating of a given region once, rather than stopping and then deciding whether it is necessary to extend the heat application. This motivates accuracy in the visual monitoring of the proximity of the two surfaces. The research problem of central interest here is to identify and address the foreseeable challenges affecting this visualization task.

D. MR Thermometry Progress

In 1982, when human MR imaging had been conducted for at least two years, image acquisition times ranged between 1 and 30 minutes, depending on the acquisition technique[9]. A review by Quesson et al[2] describes that temperature mapping by MRI was first reported in 1983 by Parker et al[10]. In 1985 Hasse et al[11] invented a new sequence for MRI acquisition, called Fast Low Angle Shot (FLASH). FLASH is considered to have reduced acquisition times by a factor of 100, compared to what was available at the time[12].

New techniques have continued to improve the refresh rate of MR thermometry. Multiple recent studies have demonstrated MR thermometry acquisition times of one second or less for images of the approximate size used in hyperthermia[13, 14].

Acquisition times are proportional to the number of voxels scanned, and Yuan et al[15] point out that reduced field of view scanning can reduce the acquisition times. Uecker et al[12] list a number of factors affecting progress in MR acquisition times over the last two decades, including improvements to: magnetic fields and coils; parallel algorithms; spiral and radial encoding; and, mathematical methods beyond the Fourier transform.

Parallel computing can reduce the processing time for some acquisition methods, but Uecker et al[12] stated that (as of 2012), while some MR vendors supported the use of a single GPU, none supported multiple GPUs.

Sloan et al[16] describe NeuroBlate® (during the trials reported on) as refreshing thermometry at approximately eight second intervals (for three consecutive slices). Vogl et al[14] reported similar acquisition times for comparable images.

There are numerous MR acquisition sequences under research or in use. All of the necessary technologies are in place to support real time MR thermometry of hyperthermia at temporal resolutions in the order of one second for typical single images. With the addition of multi-GPU support and parallel processing by manufacturers, as well as reduced field of view imaging, thermometry updates for systems such as NeuroBlate® could be reduced to as little as 1-2 seconds.

For at least a decade, real-time MRI guidance of (experimental) surgical procedures has been seen as a promising technique. Real time magnetic resonance thermometry has continued to be an area of active research over the past few years, as with Fite et al[17]

As an example of the pace of development, the prototype system for NeuroBlate® was completed four years after Monteris Medical was formed and the first commercial use of the system occurred seven years after that. This suggests that an improved visualization might be required within as little as four years. It is improbable that more than eleven years would pass before such improved visualizations are required.

E. Human Factors

The human factors impacting surgical software systems are complex. It is simple enough to monitor two boundaries in a two-dimensional plane or slice from an MRI series. It cannot be determined with certainty, however, which slice (of the 3D data) the intersection will occur in. It is therefore necessary, using current monitoring techniques, to choose multiple slices to monitor. Although software can suggest in which planes this will most likely occur, the surgeon bears responsibility for the wellbeing of the patient. The effects of age and cognitive load on performance are particularly important in the context of monitoring therapeutic lesion formation.

Given that using medical imagery is such a visual task, and given that it is desirable for surgeons to be able to have a long career, the effects of age on vision and cognitive abilities should be considered in the design of software to be used intra-operatively.[18, 19] This is supported by Hugenholtz[20] who found that the mean age of all practicing neurosurgeons in Canada in 1996 was 48.2. Bieliauskas et al[21] also found that mean response latency increased with increasing age. (by approximately 17% between age 45 and 75) Kennedy et al[18] demonstrated that visual and cognitive performance declines with age beginning "relatively early in adult life." Trick et al[19] indicated that a person's ability to accurately track multiple objects declines, from five, by approximately 1 object per decade past age 30.

In addition to increasing age, surgery is an inherently busy task. This increases the cognitive load on users during visual tasks. Holland[22] found that busier people showed decreased static and dynamic peripheral visual acuity. This could impact the ability of surgeons to detect important details in areas outside of their central focus. The need to design interfaces for users under high cognitive loads has been recognised. Roda and Thomas[23] stated that technologies should be designed to support human attentional processes. Manning et al[24] expressed that interfaces for radiologists should eliminate as many distractions as possible.

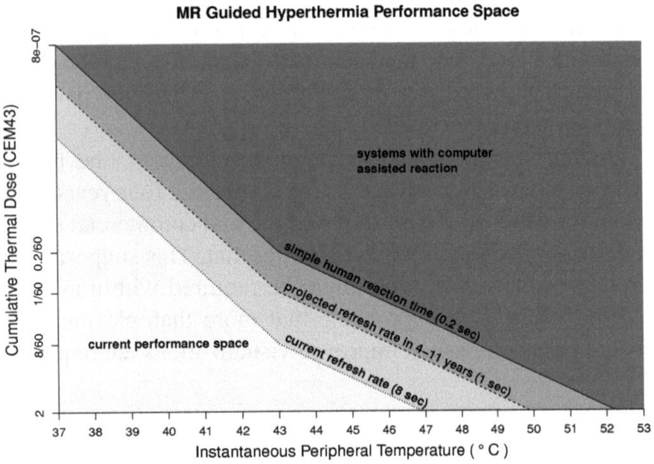

Fig. 1: Three lines, representing fixed thermometry refresh rates, are plotted using the formula for CEM43. The x-axis shows the temperature at a tumor's boundary. The y-axis (decreasing logarithmically) shows the thermal dose which accumulates at a given temperature within a fixed period of time.

II. PERFORMANCE SPACE

Monteris Medical, with its NeuroBlate® system, has demonstrated that surgeons can effectively deal with today's quality of information being updated at eight second intervals.[16] A question that we should be highly motivated to answer is: "Will surgeons be capable of dealing with higher quality data arriving at a refresh rate closer to one second, using typical current interfaces?" This question has not been addressed in the available literature. My position is that we cannot trust user interfaces to co-develop along with the medical technologies and that it will be far better to address the question proactively.

When describing the performance space within which hyperthermia systems will operate, our focus is on the temperature at the periphery of the target treatment volume. Regardless of instantaneous temperature, tissue death becomes more likely with a greater accumulated thermal dose. At higher temperatures, such as 50 °C or greater, tissue death tends to be instantaneous. The instantaneous temperatures used will depend on the particular therapy, but the accumulated thermal dose at the margins of the treatment area will be a significant concern. A safety margin of two CEM43 has previously been identified[16]. (Rhoon et al[25] suggests that actual safety margins might be as high as 2 - 9 CEM43.)

Figure 1 illustrates the performance space in which this problem exists. The x-axis shows the temperature at the periphery of a tumor. It increases, from left to right, from 37 °C (which is a normal body temperature) to 53 °C (which is higher than would typically be desirable at the periphery). The y-axis shows the amount of thermal dose that will accumulate at a given temperature and a specific period of time. It is measured in terms of CEM43. (CEM43 refers to "Cumulative Equivalent Minutes at 43 °C". It is a measure of the accumulation of thermal dose that is comparable between therapies conducted at different temperature profiles. CEM43 is discussed in detail by Sapareto and Dewey[26].) The y-axis decreases from 2 CEM43 to 8×10^{-7} °C (sufficient to show the thermal dose that accumulates within a typical human reaction time).

An increasing understanding of thermal damage mechanisms might affect the safety margin (2 CEM43 in this case). It might also limit the maximum peripheral temperature as a result of limiting the maximum tolerable amount of thermal accumulation between thermometry refreshes. Such refinement would result from an increasing understanding of thermal damage mechanisms.

Three lines are plotted in the graph. Each line represents a fixed refresh rate. Given such a period of time, a thermal dose that would accumulate in that time can be plotted for any temperature value. The change in angle at 43 °C is due to the formula for calculating CEM43. Equivalent thermal dose accumulates more slowly above 43 °C.

Areas below and above the lines are shaded. The area under the eight second line represents the performance space within which current systems such as NeuroBlate® operate. The area between the eight second line and the one second line represents the performance space that systems will likely be required to operate in within the next four to eleven years. The area between the one second line and the 200ms line represents a performance space that systems may need to operate within, but for which a timeframe has not been determined. The area above the 200ms line represents a performance space that would likely require significant automatic assistance to operate within because a human surgeon would not be able to react quickly enough to such information.

As an example of how this figure may be used: If the system delivers thermal dose refreshes at eight-second intervals and the surgeon desires to apply heat at plus or minus 2 CEM43, then the maximum peripheral temperature fitting within these parameters would be approximately 47 °C.

With time, the lowest line (current refresh rate) can be expected to move upwards. The issue is: How closely can the current refresh rate approach one second (the middle line), before the surgeon starts to have difficulty dealing with that rate of information flow using current interfaces?

A longer term issue is: If MR technology and our understanding of thermal damage mechanisms progress sufficiently to motivate systems refreshing thermometry at less than typical human reaction time, how must we build software systems to allow surgeons to deal with that?

III. Conclusion

There continues to be interest and progress in various fields relating to real-time MR guided thermal therapies. Real-time MR guided hyperthermia (such as NeuroBlate®) has not only been the subject of significant research, but has also been commercially implemented. Current human-computer interfaces support the existing technology adequately, but will become increasingly inadequate as MR capabilities, and our understanding of thermal damage mechanisms becomes more refined.

In order to maximally utilise medical and technological advancements within the next four to eleven years, improved interactive visualization will be required. Human computer interaction researchers should be motivated to investigate such improved interfaces before they are required. My future research will involve prototyping and investigating such improved interactive visualisations.

Conflict of Interest

The author declares that he has no conflict of interest.

Acknowledgments

I would like to acknowledge Dr. Henry Gardner, Dr. Bruce Thomas, Dr. Duncan Stevenson, and Dr. Hongdong Li for numerous helpful conversations.

References

1. Hassenbusch Samuel J., Pillay Prem K., Barnett Gene H.. Radiofrequency Cingulotomy for Intractable Cancer Pain Using Stereotaxis Guided by Magnetic Resonance Imaging *Neurosurgery.* 1990;27.
2. Quesson Bruno, Zwart Jacco A., Moonen Chrit T.W.. Magnetic resonance temperature imaging for guidance of thermotherapy *Journal of Magnetic Resonance Imaging.* 2000;12:525–533.
3. Sherman Jonathan, Hoes Kathryn, Marcus Joshua, Komotar Ricardo, Brennan Cameron, Gutin Philip. Neurosurgery for Brain Tumors: Update on Recent Technical Advances *Current Neurology and Neuroscience Reports.* 2011;11:313-319. 10.1007/s11910-011-0188-9.
4. Carpentier Alexandre, Itzcovitz Julian, Payen Didier, et al. Real-Time Magnetic Resonance-Guided Laser Thermal Therapy for Focal Metastatic Brain Tumors *Neurosurgery.* 2008;63.
5. Atkinson E.R.. Hyperthermia Dose Definition in *Microwave Symposium Digest, 1977 IEEE MTT-S International*:251-251 1977.
6. Milleron R.S., Bratton S.B.. 'Heated' Debates in Apoptosis *Cellular and Molecular Life Sciences.* 2007;64:2329-2333.
7. Skinner Matthew G, Iizuka Megumi N, Kolios Michael C, Sherar Michael D. A theoretical comparison of energy sources - microwave, ultrasound and laser - for interstitial thermal therapy *Physics in Medicine and Biology.* 1998;43:3535.
8. Tyc Richard, Wilson Kurt Jeff. LASER SURGERY/CANCER TREATMENT: Real-time interactivity enhances interstitial brain tumor therapy *BioOptics World.* 2010;3:AutoLITT.
9. Davis Peter L, Crooks Lawrence E, Margulis Alexander R, Kaufman Leon. Nuclear magnetic resonance imaging: Current capabilities *Western Journal of Medicine.* 1982;137:290.
10. Parker D L, Smith V, Sheldon P, Crooks L E, Fussell L. Temperature distribution measurements in two-dimensional NMR imaging. *Med Phys.* 1983;10:321–325.
11. Haase A, Frahm J, Matthaei D, Hanicke W, Merboldt K.-D. {FLASH} imaging. Rapid {NMR} imaging using low flip-angle pulses *Journal of Magnetic Resonance (1969).* 1986;67:258 - 266.
12. Uecker Martin, Zhang Shuo, Voit Dirk, Merboldt Klaus-Dietmar, Frahm Jens. Real-time MRI: recent advances using radial FLASH *Imaging.* 2012;4:461–476.
13. Stafford R Jason, Hazle John D, Glover Gary H. Monitoring of high-intensity focused ultrasound-induced temperature changes in vitro using an interleaved spiral acquisition *Magnetic resonance in medicine.* 2000;43:909–912.
14. Vogl T.J., Huebner F., Naguib N.N.N., et al. MR-based thermometry of laser induced thermotherapy: temperature accuracy and temporal resolution in vitro at 0.2 and 1.5 T magnetic field strengths. *Lasers Surg Med.* 2012;44:257-65.
15. Yuan Jing, Mei Chang-Sheng, Panych Lawrence P., McDannold Nathan J., Madore Bruno. Towards fast and accurate temperature mapping with proton resonance frequency-based MR thermometry *Quantitative Imaging in Medicine and Surgery.* 2012;2.
16. Sloan Andrew E, Ahluwalia Manmeet S, Valerio-Pascua Jose, et al. Results of the NeuroBlate System first-in-humans Phase I clinical trial for recurrent glioblastoma: clinical article. *J Neurosurg.* 2013;118:1202–1219.
17. Fite Brett Z., Liu Yu, Kruse Dustin E., et al. Magnetic Resonance Thermometry at 7T for Real-Time Monitoring and Correction of Ultrasound Induced Mild Hyperthermia *PLoS ONE.* 2012;7.
18. Kennedy Graeme J., Tripathy Srimant P., Barrett Brendan T.. Early age-related decline in the effective number of trajectories tracked in adult human vision *Journal of Vision.* 2009;9.
19. Trick Lana, Hollinsworth Heather, Brodeur Darlene A. Multiple-object tracking across the lifespan: Do different factors contribute to diminished performance in different age groups? in *Computation, cognition, and Pylyshyn* (Dedrick Don, Trick Lana. , eds.)ch. 3, :79-99MIT press 2009.
20. Hugenholtz H.. Neurosurgery Workforce in Canada, 1996 to 2011 *CMAJ.* 1996;155:39-48.
21. Bieliauskas Linas A., Langenecker Scott, Graver Christopher, Lee H. Jin, O'Neill Jillian, Greenfield Lazar J.. Cognitive Changes and Retirement among Senior Surgeons (CCRASS): Results from the CCRASS Study *Journal of the American College of Surgeons.* 2008;207:69–78.
22. Holland Dwight. *Peripheral Dynamic Visual Acuity Under Randomized Tracking Task Difficulty, Target Velocities, and Direction of Target Presentation.* PhD thesisVirginia Polytechnic Institute and State University 2001.
23. Roda Claudia, Thomas Julie. Attention aware systems: Theories, applications, and research agenda *Computers in Human Behavior.* 2006;22:557 - 587. Attention aware systems - Special issue: Attention aware systems.
24. Manning D J, Gale A, Krupinski E A. Perception research in medical imaging *Br J Radiol.* 2005;78:683-685.
25. Rhoon Gerard C, Samaras Theodoros, Yarmolenko Pavel S, Dewhirst Mark W, Neufeld Esra, Kuster Niels. CEM43 degrees C thermal dose thresholds: a potential guide for magnetic resonance radiofrequency exposure levels? *Eur Radiol.* 2013;23:2215–2227.
26. Sapareto Stephen A., Dewey William C.. Thermal dose determination in cancer therapy *International Journal of Radiation Oncology*Biology*Physics.* 1984;10:787 - 800.

Development of a Bio-Impedance Signal Simulator on the Basis of the Regression Based Model of the Cardiac and Respiratory Impedance Signals

Y.M. Mughal[1], Y.L. Moullec[1], P. Annus[1,2], and M. Min[1,2]

[1] Thomas Johann Seebeck Department of Electronics, Tallinn University of Technology (TUT), Tallinn, Estonia
[2] Competence Centre in Electronics-, Info- and Communication Technologies (ELIKO), Tallinn, Estonia

Abstract—**A software implemented bio-impedance signal simulator (BISS) is proposed, which can imitate real bio-impedance phenomena for analyzing the performance of various signal processing methods and algorithms. The underlying mathematical models are built by means of a curve-fitting regression method. Three mathematical models were compared polynomial, Fourier series and sum of sine waves with four different measured impedance cardiography (ICG) datasets and two clean ICG and impedance respirography (IRG) datasets were taken as the basis of the signals. Statistical analysis (sum of squares error, correlation and execution time) implies that Fourier series is best suited. The models of the ICG and IRG signals are integrated into the proposed simulator.**

In the simulator the correlation between heart rate and respiration rate are taken into account by means of ratio between them (5:1 respectively).

Keywords— **Regression based model, Signal Simulation and Modeling, Electrical Bio-Impedance, Impedance Cardiography, Respiratory Signal.**

I. .Introduction

Impedance cardiography (ICG) measurement has been offered as a cost effective and noninvasive method for monitoring haemodynemical parameters. The time variant part of the bioimpedance (BI) phasor reflects processes in patient physiological state since some changes in BI can be caused by normal activity or pathological reasons [1, 2].

Extracting information from impedance signals for diagnosing diseases and assessing heart function is essential for exploiting this method.

Working on real signals can be difficult; it is desirable to provide a simulation tool to enable simulation and control of such signals for analyzing the performance of various signal processing methods such as cardiac and respiratory separation algorithms, e.g. independent component analysis (ICA), adaptive filtering, ensemble averaging, and spectral methods [3, 4].

Modeling of the ICG signal has captured the interest of several researchers in the past few years, using different approaches such as described in [4, 5, 6].

In [4], Krivoshei proposed a simple bio-impedance signal synthesizer to generate cardiac and respiratory signals. The author used a piece-wise linear triangular function to model the cardiac signal and a trapezium to model the respiratory signal. The model, however, is too simple to fully imitate the cardiac and respiratory signals, and thus does not allow testing e.g. separation algorithms.

Kersulyte et al. [5] proposed a cardio model based on the sum of exponential functions. The purpose of their work was to find out a model for cardio signals as precise as possible and compare complexity parameters of the real signals and that of the model for both healthy and sick persons. They compared two function types polynomial and sum of exponentials. Their results indicate that both methods lead to similar results in terms of fidelity; however, the authors also indicate that the polynomial equation depends on the signal length and number of intervals, which could lead to too many coefficients and increased computational requirements for complex signals.

In [6] Matušek et al. proposed a cardiac signal model based on a series of real signals. By filtering and averaging the series of real signals, they estimated one average ICG signal cycle and simply replicated this cycle over time to get the final signal model. One limitation of this approach is that it lacks a mathematical model and thus the user cannot easily change the parameters of the model.

Given the limitations of the above works, it was decided to compare the suitability of three mathematical models (polynomial, Fourier series, sum of sine waves) by means of Matlab's Curve Fitting Toolbox.

II. Modeling the ICG and IRG signals

The impedance cardiography and impedance respirography (IRG) signals are nearly periodic signals that can be approximated through various mathematical models. In this study, first were evaluated ten models, which are available in the toolbox and found out that three of these gave the better results, namely polynomial, Fourier series, and sum of sine waves. Then these three models were applied on four measured ICG datasets and two clean ICG and IRG datasets for evaluation and comparison purposes. What follows briefly describes the electrical bio-impedance (EBI) measurement procedure and then discusses each model separately.

A. EBI Measurement Procedure

The datasets were obtained using multiple pairs of electrodes with different electrode configurations. The EBI measurement electrode setup is shown in Figure 1.

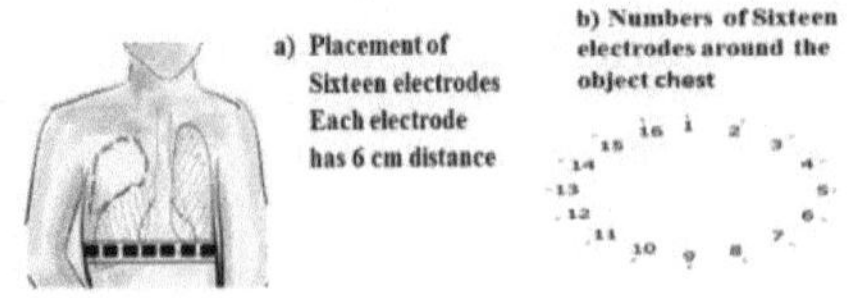

Fig 1. Sixteen electrodes configured belt, which is used for the EBI measurement procedure [2].

H. Mindedal and M. Persson (eds.), *16th Nordic-Baltic Conference on Biomedical Engineering*,
IFMBE Proceedings 48, DOI: 10.1007/978-3-319-12967-9_25

Such type of electrodes' setup is presumed to allow raising strong enough variations of the EBI in order to record the cardiac and respiration signals, which are caused by the heart and lungs. Further details about the EBI measurement setup can be found in [2].

The measured datasets are obtained from a healthy male subject aged between 40 and 50 years, in a seated position.

The total EBI dataset was divided into three different segments. Each segment contains 10 seconds of the total EBI raw data, about 10,000 samples. Accordingly, the structure of the three segments is as follows:

a) cardiac only (breathing was held),

b) cardiac + respiration (deep breathing),

c) cardiac + respiration + motion artefacts (normal breathing with added motion artefacts).

In what follows, the four ICG datasets correspond to b) and the clean ICG and IRG datasets correspond to filtered versions of a) and b), respectively.

B. Models and Evaluation Method

a) Polynomial Model

Polynomials are well suited for cases where a fairly simple empirical model is needed; they can be used for interpolation or extrapolation to characterize data by means of a global fit. The general polynomial model formula is given in Equation 1:

$$y = \sum_{i=1}^{n+1} p_i t^{n+1-i} \quad (1)$$

where n is the degree of the polynomial (highest power of the predictor variable), $n+1$ is the order of the polynomial (number of coefficients), p_i are the coefficients and t is time.

In this work, the polynomial model was evaluated for degrees 1 to 9 for the different datasets; degree 9, which is the highest order available in the toolbox, gave the best suitable results. The comparative results are shown in Table 1 and Figures 2 & 3.

b) Fourier Series Model

The Fourier series is a sum of sine and cosine functions that describes a periodic signal. The model formula is given in Equation 2:

$$y = a_0 + \sum_{i=1}^{n} a_i \cos(i\omega t) + b_i \sin(i\omega t) \quad (2)$$

where a_0 is the intercept, which is constant term in the data, ω is the fundamental frequency and n is the number of terms in the series. The model was evaluated with 1 to 8 terms for the different ICG datasets; the best suitable results were obtained for the degree of 8, the highest available in the toolbox. The comparative results are shown in Table 1 and Figures 2 and 3.

c) Sum of Sine Waves Model

This model consists of a sum of sine terms only. The model formula is given in Equation 3:

$$y = \sum_{i=1}^{n} a_i \sin(i\omega t + c_i) \quad (3)$$

where a is the amplitude, ω is the frequency, c the phase, which is constant for each term and n is the total terms in series.

The model was evaluated with 1 to 8 terms for the different datasets; 8 terms (the highest available in the toolbox) gave the most suitable results. The comparative results are shown in Table 1 and Figures 2 and 3.

d) IRG Signal with Polynomial, Fourier Series and Sum of Sine Waves Models

Following the same approach as for the ICG signal, the IRG clean dataset is also modeled with the polynomial, Fourier series and sum of sine waves methods. The comparative results are shown in Table 1 (Clean IRG) and Figure 3(c).

C. Statistical Parameters

The performance of the three modeling methods is evaluated by means of the following fit measures.

a) Sum of Squares Error (SSE)

The SSE statistic assesses the total deviation of the data values from the fitted model, as expressed in Equation 4:

$$SSE = \sum_{i=1}^{n} w_i \,(y_i - \overline{y}_i)^2 \quad (4)$$

where n is the number of data points, y_i is the response data, and $\overline{y}_i$ is predictor data. SSE values close to 0 indicate that the model is fitted well and has a very small random error [7].

b) R-Square

R-Square measure is the square of the correlation between the data and the fitted model values. A value close to 1 shows a greater correlation between the data and the model whereas a value close to 0 shows a poor correlation. It is determined as the ratio of the sum of squares of the regression (SSR) and the total sum of squares (SST), where $SST = SSR + SSE$. The R-square measure is given in Equation 5 [7]:

$$R\text{-}square = \frac{SSR}{SST} = 1 - \frac{SSE}{SST} \quad (5)$$

c) Execution time

The execution time is measured through Matlab stopwatch functions (tic, toc) and reported in Table 1.

III. EXPERIMENTAL RESULTS

Table 1 and Figures 2 and 3 show the fit of the three models with the various datasets. Generally speaking, the three models provide a reasonable fit across the four datasets: the average SSE value is 0.879e-07, the min and max values are 0.161e-07 and 1.9417e-07, respectively

Similarly, the average R-square value across the four datasets is 0.9762, the min and max values are 0.9512 and 0.9936, respectively.

The Fourier series model minimizes the error (average SSE=0.335e-07) and has also a high correlation across the four datasets as compared to the other models. However, it took 1.275 more seconds to execute as compared to the polynomial model; it is nevertheless much faster (by 44.476 seconds or nearly 10 times) than the sum of sine waves model.

In this study, the most suitable results were obtained with eight terms for the Fourier series model, which gives 18 coefficients. For the polynomial model, we set the degree to 9, leading to ten coefficients. It is preferable to limit the number of coefficients for relating them to the patients' condition. However, this has to be traded-off for a lower fit, as shown in Table 1.

Table 1. Evaluation Criterions Results for the Modeled Signal

Datasets	Sum of sine Waves (24 coeff)		Fourier (18 coeff)		Polynomial (10 coeff)					
	SSE	*R-Sq*	*SSE*	*R-Sq*	*SSE*	*R-Sq*	*SSE Avg*	*SSE Min*	*SSE Max*	*R-SqAvg*
Dataset 1	1.0424e-07	0.9917	0.1612e07	0.9987	1.2270e-07	0.9903	0.810e-07	0.161e-07	1.23e-07	0.9935
Dataset 2	0.9044e-07	0.9875	0.1786e-07	0.9976	0.3050e-07	0.9959	0.463e-07	0.179e-07	0.904e-07	0.9936
Dataset 3	1.9417e-07	0.9274	0.6476e-07	0.9758	1.3185e-07	0.9506	1.326e-07	0.6476e-07	1.9417e-07	0.9512
Dataset 4	0.8054e-07	0.9714	0.3506e-07	0.9876	1.6683e-07	0.9409	0.941e-07	0.3506e-07	1.6683e-07	0.9666
SSE Avg, R-Sq Avg	1.17e-07	0.970	0.335e-07	0.9758	1.13e-07	0.969	0.879e-07			0.9762
SSE Min, R-Sq Min	8.05e-08	0.161	0.161e-07	0.9758	0.305e-07	0.941		0.161e-07		0.9512
SSE Max, R-Sq Max	1.94e-07	0.9917	0.648e-07	0.9987	1.67e-07	0.996			1.9417e-07	0.9936
Clean ICG Signal with different scale										
Clean ICG	0.1996	0.9994	0.0611	0.9999	2.8229	0.9937	1.0279	0.0611	2.8229	0.9959
Ex. Time (s)	~49.170		~4.694		~3.419					
Clean IRG Signal with different scale										
Clean IRG	7896.1e-07	1	2890.6e-07	1	19.5782	0.9983	6.5264	2890.6e-07	19.5782	0.9994

Regarding the difference between the polynomial and the sum of sine waves models, it can be seen that for Datasets 2 and 3, the polynomial model minimizes the error (0.3050e-07 and 1.3185e-07, respectively) and is highly correlated with the datasets (0.9959 and 0.9506, respectively). On Datasets 1 and 4, the sum of sine waves model minimizes the error (1.0424e-07 and 0.8054e-07, respectively) and is highly correlated (0.9917 and 0.9714 respectively) with the datasets. However, 8 terms were used for the sum of sine waves model, which gives 24 coefficients (versus 10 for the polynomial model) and a much longer execution time.

For the clean ICG and IRG datasets, the Fourier series model performed very well among all to minimize the error (0.0611 and 2890.6e-07, respectively) and is highly correlated (0.9999 and 1, respectively) with the datasets. It is followed by the sum of sine waves model, which has the second minimum error (0.1996 and 7896.1e-07, respectively) and high correlation (0.9994 and 1, respectively) but also has a larger number of coefficients (24) and larger execution time (49.170 seconds) as compared to the polynomial model.

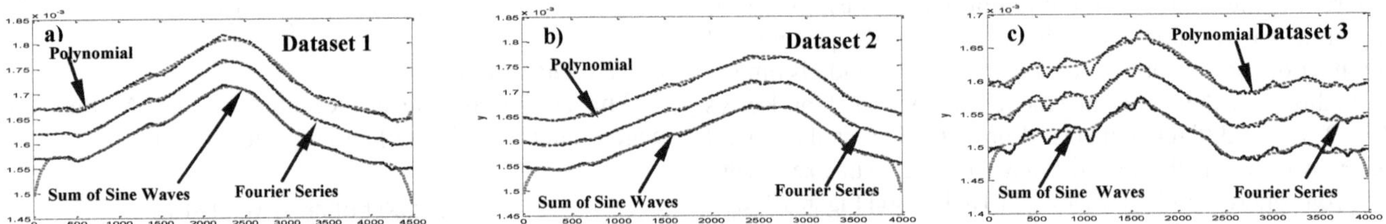

Fig 2. Measured datasets (solid-lines) and fitted models (dotted-lines) for three EBI datasets: a) results of fitting of the EBI dataset 1, b) results of fitting of the EBI dataset 2, c) results of fitting of the EBI dataset 3. Results for the sum of sine waves model are presented without offset, results for Fourier series model are offset by 0.05×10^{-3} and results for Polynomial model are offset by 0.1×10^{-3}.

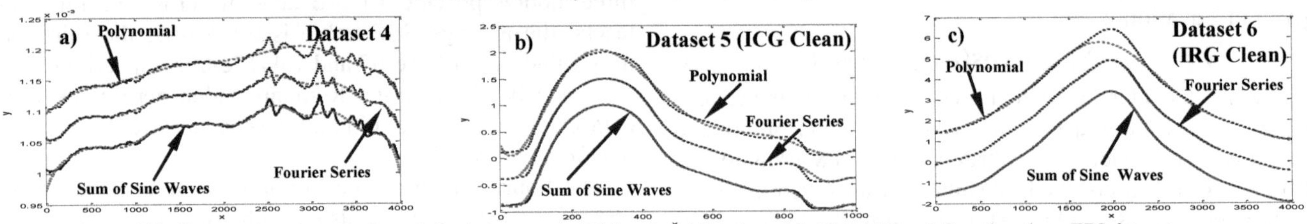

Fig 3. Measured (a) and cleaned (b, c) datasets (solid-lines) and fitted models (dotted-lines) for other three EBI datasets: a) results of fitting of the EBI dataset 4, b) results of fitting of the cleaned ICG dataset 5, c) results of fitting of the cleaned IRG dataset 6. Results for the sum of sine waves model are presented without offset, results for Fourier series model are offset by offset a) 0.05×10^{-3}, b) offset 0.5, c) offset 1.5) and results for Polynomial model are offset by (a) 0.1×10^{-3}, b) 1, c) 3].

IV. THE BIOIMPEDANCE SIGNAL SIMULATOR (BISS)

This section describes how the Fourier series model was included in our Bioimpedance Signal Simulator (BISS).

As shown in Figure 4, the simulated bio-impedance signal is generated by summing the ICG signal ($S_{\Delta Z\ ICG}$), artefacts ($S_{Artefacts}$), a white Gaussian noise (S_{Noise}) and the IRG signal ($S_{\Delta Z\ IRG}$) such as:

$$S_{EBI(t)} = S_{\Delta Z\ ICG} + S_{Artefacts} + S_{Noise} + S_{\Delta Z\ IRG} \quad (6)$$

The BISS' GUI is shown in Figure 5, where a) is the menu used to perform different operations such as loading different datasets (ICG/FCG) to simulate the signal, saving the final generated EBI signal model for further processing and exiting from the BISS environment, b) a recorded clean ICG period, c) a period of the ICG signal model, d) a recorded respiration period e) a period of the IRG signal model f) the continuously simulated ICG signal.

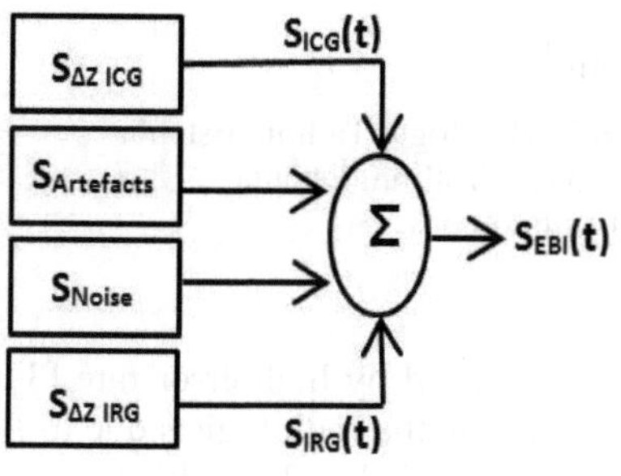

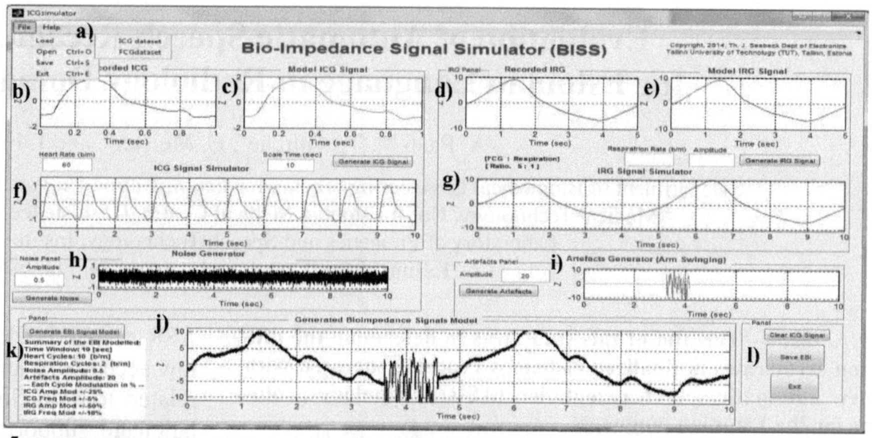

Fig 5. User Interface of the Bioimpedance Signal Simulator (BISS).
The heart rate, time scale, respiration rate, noise and atefacts amplitude parameters are user-Controlled

In order to take the real phenomena of BI signals into account, a random modulation is introduced with each cycle (amplitude ±25, frequency ±5). Moreover, the user should specify the heart rate in beats/min and time window. g) is the continuously simulated respiration signal where a random modulation is introduced with each cycle (amplitude ±50, frequency ±10).

The respiration rate is correlated to the cardiac heart rate by means of the ratio. The default ratio is 5:1 (5 cardiac cycles for 1 respiration cycle). Nevertheless, the user can control the respiration rate as well. h) is the noise generator, i) the recorded artefacts caused by swinging the arm during the measurement (randomly moving in the defined time window, j) the generated bio-impedance signal model based on the user entered parameters, k) the detailed summary of the generated bio-impedance signal model and l) buttons that let the user clear all simulated model signals and start again, save the EBI signal model and exit from BISS' GUI environment.

Figures 5 f), g), h) and i) illustrate the effect of the user-controlled parameters such as time scale window, heart rate (b/m), respiration rate (b/m), noise amplitude and artefacts amplitude.

V. Conclusions

The polynomial model is relatively simple, but it does not provide the best results for our application. The sum of sine waves model produces better results than the polynomial one, but is less suitable than the Fourier series one because it has a higher number of coefficients, higher SSE values, lower R-Square values, and higher execution times.

Overall, the Fourier series model fits with the measured datasets very well, minimizes the error and has high correlation values as compared to the two other models; only its execution time is slightly higher than that of the polynomial model.

Furthermore, the correlation between the heart rate and the respiration rate is implemented by means of a ratio (default 5 ICG cycles for 1 IRG cycle).

Finally, the user can enable the insertion of the recorded artifact in the final EBI model.

Nevertheless, the resulting simulated signal does not model all aspects of the real bioimpedance data yet. Thus, future work will refine the model by means of piece-wise segmentation of the datasets for finer grain curve-fitting while maintaining the number of coefficients to the required minimum for reflecting the pathological conditions (i.e. not necessarily 24, 18, and 10 as shown in Table 1).

The Starling's and Poiseuille laws will be taken into account in the model to reflect the systolic and diastolic phenomena respectively.

Acknowledgment

The authors thanks, Prof. T. Rang, Dr. T. Parve and Dr. A Krivoshei, for providing valuable advices.

Research supported by the European Union through the European Regional Development Fund in frames of the research center CEBE, the competence center ELIKO, the ESF DoRa, the Estonian Ministry of Education and Research (Institutional Research Project IUT19-11) and Tiger University.

References

1. Grimnes S .and Martinsen. Ø. G. (2008). Bioimpedance & Bioelectricity Basics. London: Academic Press.
2. Mughal Y. M. (2014) Decomposing of Cardiac and Respiratory Signals from Electrical Bio-impedance Data Using Filtering Method The Int. Conf. on Health Inf. IFMBE Proc. Vol. 42, pp 252-255
3. Mughal Yar M., Krivoshei A, Annus P. (2013) Separation of cardiac and respiratory components from the electrical bio-impedance signal using PCA and fast ICA Int. Conf. on Control, Engineering & Information Technology, Proc. Eng. & Tech., Vol.1,
4. Krivoshei A. (2006) A Bio-Impedance Signal Synthesiser (BISS) for Testing of an Adaptive Filtering System", Proc, of the BEC . p. 1-4.
5. Kersulyte G, Navickas Z, Raudonis V (2009) Investigation of Complexity of Extraction Accuracy Modeling Cardio Signals in Two Ways IEEE Int. Workshop on Intelligent Data Acquisition and Advanced Computing Systems: Tech. and Appl. 21-23.
6. Matušek A, (2012) Modelling of Impedance cardiac Signals M.S. thesis, Faculty of Electrical Eng and Comm. Dept. of Biomedical Eng, Brno university of Technology, Czech Republic.
7. Matlab (2012b) manual, Curve Fitting Toolbox, (June 15, 2014).

Author: Y. M. Mughal
Institute: Thomas Johann Seebeck, Department of Electronics, TUT
Street: Ehitajate tee 5
City: Tallinn
Country: Estonia
Email: yar@elin.ttu.ee.

Evaluation of Automatic Speech Recognition Prototype for Estonian Language in Radiology Domain: A Pilot Study

A. Paats[1,2], T. Alumäe[3], E. Meister[3], and I. Fridolin[1]

[1] Department of Biomedical Engineering, Technomedicum, Tallinn University of Technology, Tallinn, Estonia
[2] Medical Technology, North Estonian Medical Centre, J. Sütiste tee 19, 13419, Tallinn, Estonia
[3] Laboratory of Phonetics and Speech Technology, Institute of Cybernetics, Tallinn University of Technology, Tallinn, Estonia

Abstract — **The aim of this study was to determine the dictation error rates in finalized radiology reports generated with a new automatic speech recognition (ASR) technology prototype for the Estonian language.**

For training a language model, 177 659 real radiology reports from different imaging modalities were used. Manually normalized versions of 1299 randomly selected reports were created to standardize the report corpus. The ASR prototype, incorporating the trained language and acoustic models, was tested in Radiology Department, North Estonia Medical Centre, Tallinn, Estonia, by 17 radiologists (11 female and 6 male). In total, 424 reports were dictated, including 77 067 x-ray, 30 929 ultrasound, 28 825 computed tomography, 14 815 mammography, 12 082 endoscopic, 8 792 magnetic resonance tomography, 3 950 radiology consultation and 1 199 angiographic reports. Word error rates (WER) and report error rates (RER) were calculated for each speaker and modality.

Total WER over all material was 18.4% and total RER 93.1%. WER and RER were lowest for mammography dictations (7.7%; 70.3%), and highest for angiography (34.4%; 100%), followed by endoscopy (30.9%; 100%). 3D modalities had higher RER and WER compared to planar x-ray correlating with the complexity of the radiology reports. Live experiments with the ASR prototype showed differences between the users depending on their experience and speech characteristics.

In summary, the ASR prototype for Estonian language in radiology domain was the first time successfully applied and assessed in routine clinical practice. Improvements of the ASR prototype performance are planned in the future.

Keywords— **automatic speech recognition, radiology, Estonian language, reporting, word error rate**

I. Introduction

In speech recognition, dictated speech is converted to digital signal and then to a sequence of words in written text [1]. Automatic speech recognition (ASR) technology has dramatically improved over the past several years, and there are several commercialized applications available. The benefits are improved patient care and resource management in the form of reduced report turnaround times, reduced staffing needs, and the efficient completion and distribution of reports [2]. However, effective utilization of ASR system could be hampered by high error rate [3], [4], low acceptance and interest by the radiologists due to issues related to workflow or culture [5], [6]. Lack of a mother-language supported ASR system for under-resourced and agglutinative languages could be one reason [7]. Apart a preliminary attempt [8], no Estonian language based ASR systems exist currently in radiology.

The scientists from Tallinn University of Technology in collaboration with radiologists from North-Estonian Medical Centre (NEMC), Tallinn, Estonia, took a step closer towards an ASR application in radiology for Estonian language by performing a study using Estonian based models. Since ASR technology in its development phase has a high frequency of transcription errors, necessitating careful proofreading and report editing, a profound understanding about the errors and the frequency of errors is inevitable.

The aim of this study was to determine the dictation errors in finalized radiology reports generated with ASR technology prototype for Estonian.

II. Materials and Methods

An ASR system is based on three models: acoustic model (AM), language model (LM) and a pronunciation lexicon (PL). AM describes the spectral and temporal characteristics of individual phonemes in different contexts. In most modern ASR systems, hidden Markov models are used for representing context-dependent phonemes. AMs are trained on large human-transcribed speech corpora where origin does not have to exactly match the domain of the ASR use.

LM lists the words of the language and describes how they are statistically combined. Statistical n-gram language models are trained on large text corpora that must match the language of ASR application as closely as possible

Pronunciation lexicon links the AM and LM by the mapping of words in the LM to sequences of AM units.

A. Text corpus

Language models are based on the text corpus of specific language and domain of usage. For preparation the text

H. Mindedal and M. Persson (eds.), *16th Nordic-Baltic Conference on Biomedical Engineering*,
IFMBE Proceedings 48, DOI: 10.1007/978-3-319-12967-9_26

corpus of Estonian language in radiology domain the real radiology reports were used. The 177 659 reports, interpreted during one year from May 1st, 2012 to April 30th, 2013, were retrieved from Radiology Information System of NEMC. All reports were anonymized and patient specific data was removed. The proportion of each modality was calculated inside of full set of reports. It included 77 067 x-ray, 30 929 ultrasound, 28 825 computed tomography, 14 815 mammography, 12 082 endoscopic, 8 792 magnetic resonance tomography, 3 950 radiology consultation and 1 199 angiographic reports.

The reports in the daily clinical practice have been created by radiologists themselves via a computer keyboard. To save time during typing, abbreviations for various medical terms are used in the reports very frequently. Due to the fact that every radiologist has his/her own style of writing and abbreviation, there is a lot of variety to represent the same concept.

In order to standardize the report corpus, manually normalized versions of the 1299 randomly selected reports were created. During this process the common understanding for medical abbreviations, acronyms and date-time style was agreed. Every selected report was reviewed and normalized by two senior radiologists. The normalization included removing lexical, grammatical, terminological and other errors, expanding unnecessary abbreviations, homogenizing dates and times.

To assist ASR pronunciation model development, the manually normalized reports were also supplied with pronunciation information. All unconventionally pronounced words, unexpanded abbreviations and acronyms were transliterated to their spoken form.

Finally, the three versions of the reports (original, normalized and pronunciation) were compiled into a parallel corpus for developing ASR LM and PL.

B. Language model training and text corpus processing

The hybrid model was used to create automatically spoken form transliterations for all other 177 659 reports in the corpus.

First, manual rules for normalizing the unnecessary abbreviations in the original reports were constructed and n-gram statistics was created to select between the several competing normalization for various inflections and the disambiguation model was trained.

In a similar way we created a model for transforming normalized reports into their spoken form transliterations.

By concatenating the two transformation models, we converted all the reports in our corpus to transliterated spoken form. The resulting corpus of about 10 million words served as the training data for the statistical LM for speech recognition. The LM vocabulary was created by selecting all words from the corpus that occurred at least twice. This resulted in 52 297 words.

C. Description of Automatic Speech Recognition System prototype and implementation for live experiments

The ASR system consists of a server component that takes care of decoding speech and normalizing the recognized hypotheses and a client component that is responsible for recording speech and presenting the recognition results to the user.

The server component consists of two parts: a master server and a worker pool. The master server forwards client's audio to the workers and sends the results submitted by the workers back to the client. Worker handles actual speech decoding using the Kaldi toolkit [9]. Workers can be dynamically started and stopped on remote servers, making it possible to handle a very high number of parallel recording sessions.

The client component is implemented as Java application that communicates with the server using a protocol based on websockets.

The AMs for Estonian language are trained on approximately 135 hours of speech from various non-medical sources. Speaker independent discriminatively trained triphone Gaussian mixture models are used.

D. Testing of ASR in real clinical environment

ASR prototype was tested in Radiology Department of NEMC. Radiologist standard workplace consists of PC equipped with 4 monitors. One monitor is used for composing of report in Radiology Information System and others for visualization of images with PACS (Picture Archiving and Communication System) client (Agfa, Impax 6.4). A web interface of the ASR prototype was implemented into the same monitor as RIS in the way, that the radiologist had visual control of both systems at the same time. Every station, where prototype was tested, was equipped with a high quality microphone headset (Logitech USB H340).

An instruction manual, describing how to use prototype and how to dictate different text components, as agreed during report normalization process, was given to test users.

The dictating radiologist's code and study accession number was stored by the prototype web interface. Speech recognition was done in real time during dictation. Every recognized sentence was checked by radiologist immediately after dictation and incorrectly recognized words or phrases were corrected. Both, the text recognized by ASR and the text corrected by radiologist, were stored by the prototype interface for future analysis. In order to be able to remember all details of dictations for long reports, the correction was done after every sentence.

Totally 17 radiologists (11 female and 6 male) were participating in the testing, among whom 12 were skilled radiologists with work experience over 5 years, and 5 were radiology residents under training. 3 radiologists had previous experience of using ASR in other languages. From all radiologists 15 were the native Estonian language speakers and two with different mother tongue but highly skilled in Estonian.

During prototype testing 424 reports were dictated. Distribution of dictated reports between radiologists and modalities is presented in Table 1.

Live experiments with the prototype showed differences between users. Some of them frequently forgot to switch on ASR before starting of dictation. Some users discovered that ASR is not recognizing specific acronyms, words, punctuation symbols or capital letters correctly and got stressed. There was also a problem with persons who have naturally very low voice intensity. For them the microphone sensitivity was tuned to maximum. Due to the variable accuracy of ASR prototype and the need to make a lot of corrections the testing of ASR prototype was taking much more time than normal reporting, and it was found to be stressing for some radiologists.

Table 1. Distribution of dictated reports between modalities (RG: X-Ray, CT: Computed Tomography, MR: Magnetic Resonance; MG: Mammography; US: Ultrasound; AG: Angiography; ES: Endoscopy)

Radiologist	RG	CT	MR	MG	US	AG	ES
#1		5	5				
#2	1	20					
#3	3	7			10		
#4		1	26				
#6	12	3	4		1		
#8							4
#9							33
#10				18	5		
#11	14	6					
#12	15	4			1		
#13	30						
#14	10	12	15		13		
#16		50	13				
#17		18	2				
#19		3	19			2	
#21	20						
#22				19			
Total	159	161	92	37	34	2	37

The dictated reports were analyzed for finding the errors of ASR system. Word error rates (WER) and report error rates (RER) were calculated, as described in [3], for each speaker and modality group.

III. Results

Total WER over all material was 18.4% and total RER 93.1%. As seen from the Table 2, the mean WER over all speakers was 17.7% (SD 8.4), and the mean RER was 92.7% (SD 14.7).

Dictations of radiologist #10 and #22 had lowest WER and dictations of radiologists #9 and #8 had highest WER. Dictations of radiologist #21 and #10 had lowest RER. Dictations of 11 radiologists had all reports with recognition errors (RER 100%).

Table 2. WER and RER by speakers

Radiologist	No of Reports	No of Words	WER	RER
#1	10	1286	18.8	100
#2	20	2428	17.2	100
#3	7	1249	22	100
#4	16	1322	29.4	100
#6	3	441	11.8	100
#8	4	220	31.8	100
#9	8	349	30.4	100
#10	18	519	7.7	66.7
#11	5	216	11.1	100
#12	1	34	11.8	100
#13	1	12	8.3	100
#14	46	2416	27.7	95.7
#16	57	6306	12.9	94.7
#17	20	1269	24	100
#19	24	1885	16.5	95.8
#21	2	44	11.4	50
#22	19	508	7.7	73.7
Mean	15.4	1206.1	17.7	92.7
SD	15.7	1537.7	8.4	14.7

WER and RER of dictations were calculated for each modality group. The results are given in Table 3.

Table 3. WER and RER by modality

Modality	No of Reports	No of Words	WER	RER
CT	119	12541	15.4	97.5
AG	2	90	34.4	100
ES	12	569	30.9	100
MG	37	1027	7.7	70.3
MR	66	5397	25.6	98.5
RG	13	333	27.6	84.6
US	12	547	13	91.7
Mean	37.3	2929.1	22.1	91.8
SD	42.1	4621.4	10.1	11

WER and RER were lowest for mammography dictations (7.7%; 70.3%), and highest for angiography (34.4%; 100%) and endoscopy (30.9 %; 100%).

IV. Discussion

This paper describes an evaluation of an ASR prototype for the Estonian language in radiology domain, including WER and RER analyses by user and modality. An important outcome was that it was the first time when the ASR prototype was used in routine clinical practice in Estonia.

Table 2 shows highest WER and RER values for radiologists #9 and #8 who reported only endoscopy studies (Table 1). Reporting in endoscopy is not well standardized and content of original reports fluctuated widely and this shows that standardized rules did not apply correctly during dictation experiments. The same table shows lowest WER and RER for the speakers who reported mostly mammography studies (#10 had 18 MG reports from the total of 23, and #22 had all MG reports). This is due to short and similar reports for MG modality. Moreover, the MG reports tend to follow a more rigorous and standardized structure [3]. Total WER over all material was 18.4% and total RER 93.1%, which is still too high for routine clinical usage.

Table 3 shows that the reports dictated for complicated 3D modalities (CT, MR) have higher WER and RER values compared to x-ray, US and MG. The results confirm the findings for MG as described comparing the users above, and also findings from an earlier study [3], according to the probability for errors was 4.4 times higher for MRI than MG. Because the number of angiography studies is small compared to the other modalities, and the number of normalized angiography reports in text corpus is relatively small, the performance of ASR for angiography reports is rather low. However, this is in concordance with the results, that the reports of non radiography modalities, including MRI and AG, tend to have higher risk of error [4]. The reasons for the low performance of ES were described above.

In summary, current WER and RER values are still insufficient to achieve shorter reporting times compared to direct keyboard typing. According to the feedback from the testing radiologists, the ASR system for Estonian language in radiology domain could be taken into usage as a daily tool, assuming that performance of ASR will be improved.

V. Conclusions

It was shown successfully that it is possible to develop non-commercial ASR prototype for Estonian language in radiology domain for routine clinical usage in Estonia.

According to the feedback from testing radiologists, the ASR system could be taken into usage as a daily tool, which enables shortened reporting and turnaround times, assuming that performance of ASR will be improved.

In the future we are planning to improve the performance of the ASR prototype by improving the handling of acronyms and abbreviations in the LM and adapting the LM and AM to individual speakers and modalities.

Acknowledgment

The authors wish to thank all radiologists who participated in the live experiments, especially Dr Roose and Dr Raudvere for normalization of the report corpus. The work is supported by the European Union through the European Regional Development Fund, project 3.2.1201.13-0010.

References

1. Koivikko MP, Kauppinen T, Ahovuo J (2008) Improvement of report workflow and productivity using speech recognitionâ€"a follow-up study. Journal of Digital Imaging 21:378-382
2. Voll K, Atkins S, Forster B (2008) Improving the utility of speech recognition through error detection. J Digit Imaging 21:371-377
3. Basma S, Lord B, Jacks LM, Rizk M, Scaranelo AM Error rates in breast imaging reports: comparison of automatic speech recognition and dictation transcription. Am Roentgen Ray Soc 197:923-927
4. Chang CA, Strahan R, Jolley D (2011) Non-clinical errors using voice recognition dictation software for radiology reports: a retrospective audit. J Digit Imaging 24:724-728
5. Talton D (2005) Perspectives on speech recognition technology. Radiology Management 27(1):38 - 40
6. Pezzullo JA, Tung GA, Rogg JM, Davis LM, Brody JM, Mayo-Smith WW (2008) Voice recognition dictation: radiologist as transcriptionist. J Digit Imaging 21:384-389
7. Arisoy E, Arslan ML (2004) Turkish Radiology Dictation System. Proceedings of SPECOM St Petersburg, Russia
8. Alumäe T, Meister E (2010) Estonian large vocabulary speech recognition system for radiology. In: Skadina I, Vasiljevs A (eds) Human Language Technologies. The Baltic Perspective: Proceedings of the Fourth International Conference, Baltic HLT 2010, vol 219. Amsterdam: IOS Press, pp 33 - 38
9. Povey, D, Ghoshal, A, Boulianne, G, Burget, L, Glembek, O, Goel, N, Hannemann M, Motlicek P, Qian Y, Schwarz P, Silovsky J, Stemmer G, Vesely, K. (2011) The Kaldi speech recognition toolkit. In: Proceedings of ASRU. Hawaii, USA.

Author: Andrus Paats
Institute: Department of Biomedical Engineering, Technomedicum, Tallinn University of Technology
Street: Ehitajate tee 5
City: 19086 Tallinn
Country: Estonia
Email: Andrus.paats@regionaalhaigla.ee

Atrial Impulses at the Atrioventricular Node: Arrival versus Conduction during Atrial Fibrillation

L. Iozzia[1], G. Garoldi[1], F. Sandberg[2], V.D.A. Corino[1], L.T. Mainardi[1], and L. Sornmo[2]

[1] Department of Electronic, Information and Bioengineering, Politecnico di Milano, Milan, Italy
[2] Department of Bioengineering, Lund University, Lund, Sweden

Abstract— **The atrioventricular (AV) node plays a central role during atrial fibrillation (AF). We have recently proposed a statistical AV node model defined by parameters characterizing the arrival rate of atrial impulses, the probability of an impulse choosing the slower of the two AV nodal pathways, the refractory periods of the pathways, and the prolongation of refractory periods. All model parameters are estimated from the RR series using maximum likelihood estimation, except for the mean arrival rate of atrial impulses which is estimated by the AF frequency derived from the f-waves. The aim of this study is to assess the relationship between the probability γ of an atrial impulse to arrive at the slow pathway with the probability α (used in the model) of an impulse to pass through that pathway. A theoretical equation was derived to relate α with γ and viceversa, highlighting the dependence on both refractory periods and their prolongation. Results based on simulations show that the equation was correct, with an average absolute error of 0.0034 ± 0.0008.**

Keywords— **Atrial fibrillation, atrioventricular node, models, concealed conduction.**

I. INTRODUCTION

During atrial fibrillation (AF), a large number of atrial impulses bombard the atrioventricular (AV) node and some of them are blocked. Even if the important role played by the AV node is widely recognized, the relationship between atrial and ventricular rates and AV node has not been thoroughly studied. Various models of the AV node during AF have been proposed, either models for simulation or models for parameter estimation involving observed data. Simulation models aim at explaining AV nodal characteristics [1] or the effect of pacing [2], [3]. Although such models can offer detailed characterization of the underlying electro-physiological dynamics, they are unsuitable for parameter estimation due to problems of uniqueness as a consequence of the large number of parameters which must be subjected to optimization. Recently, we proposed a statistical model of AV nodal function during AF which lends itself well to ECG-based parameter estimation [4, 5]. The model is defined by a small set of parameters which characterizes the arrival rate of atrial impulses, the probability of an impulse choosing either one of the dual AV nodal pathways, the refractory periods of the pathways, and the prolongation of refractory periods. The parameters were estimated from the RR series using maximum likelihood (ML) estimation, except for the mean arrival rate of atrial impulses which was estimated by the AF frequency derived from the f-waves of the ECG [6]. The aim of this study is to assess the relationship between the probability of an atrial impulse to arrive at one pathway with the probability (used in the model) of an impulse to pass through that pathway.

II. METHODS

A. Description of the AV node model

In the present model [4, 5], the AV node is treated as a lumped structure which accounts for concealed conduction, relative refractoriness, and dual AV nodal pathways. Atrial impulses are assumed to arrive to the AV node according to a Poisson process with mean arrival rate λ. Each arriving impulse is suprathreshold, i.e., the impulse results in ventricular activation unless blocked by a refractory AV node. The probability of an atrial impulse passing through the AV node depends on the time elapsed since the previous ventricular activation t. The refractory period is defined by both a deterministic part τ and a stochastic part τ_p, the latter modelling prolongation due to concealed conduction and/or relative refractoriness and assumed to be uniformly distributed over the interval $[0; \tau_p]$. Hence, all atrial impulses arriving to the AV node before the end of the refractory period τ are blocked. Then follows an interval $[\tau; \tau + \tau_p]$ with linearly increasing likelihood of penetration into the AV node. Finally, no impulses can be blocked if they arrive after the end of the maximally pro-longed refractory period $\tau + \tau_p$. The mathematical characterization of refractoriness of the *ith* pathway (i = 1; 2) is thus defined by the positive-valued function

H. Mindedal and M. Persson (eds.), *16th Nordic-Baltic Conference on Biomedical Engineering*,
IFMBE Proceedings 48, DOI: 10.1007/978-3-319-12967-9_27

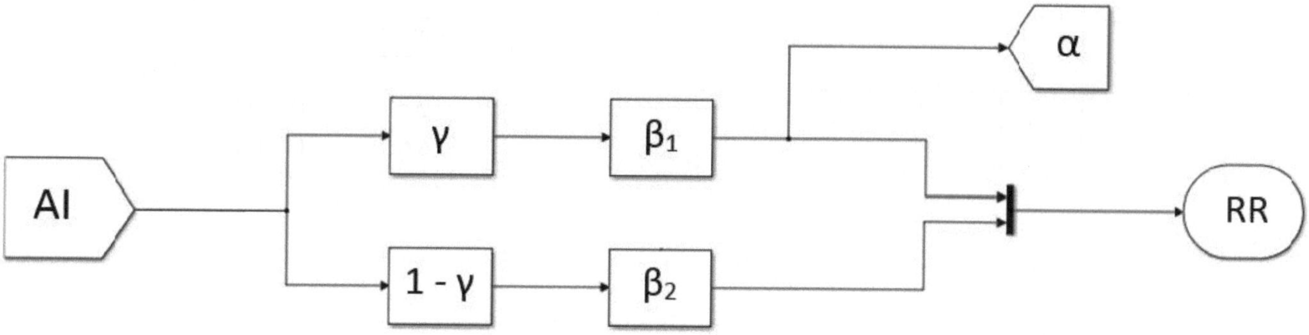

Figure 1: Block diagram of the AV node model. AI = atrial impulses; γ = probability of an impulse to arrive at the slow path; α = probability of an impulse to pass through the slow path; β_1 = prolongation of the refractory period of the slow pathway; β_2 = prolongation of the refractory period of the fast pathway.

$$\beta_i(t) = \begin{cases} 0 & \text{if } 0 < t < \tau_i \\ \dfrac{t-\tau_i}{\tau_p} & \text{if } \tau_i \leq t < \tau_i + \tau_p \\ 1 & \text{if } t \geq \tau_i + \tau_p \end{cases} \quad (1)$$

where t denotes the time elapsed since the preceding ventricular activation. In 1, the deterministic part of the refractory period is assumed to have either a length of τ_1 or τ_2, depending on the penetrating pathway. The probability of an atrial impulse to take the pathway with the shorter refractory period τ_1 is equal to α, and accordingly the other pathway is taken with probability $(1-\alpha)$.

B. Arriving vs. passing through

We empirically investigated the relationship between α (the probability of an atrial impulse to pass through the slow pathway and γ (the probability of an atrial impulse to arrive at the slow pathway). When γ is given as input, α can simply be computed as the ratio between the number of impulses passed through the slow pathway and the total number of conducted impulses (see Figure 1).

The following relation was found:

$$\alpha = 1 - \left[(1-\gamma)e^{-\lambda\gamma\Delta\tau}\right] \quad (2)$$

The parameter α depends on i) $\lambda\gamma$ and ii) $\Delta\tau = \tau_1 - \tau_2$. The term $\lambda\gamma$ is the probability with which the atrial impulses choose the slow pathway. Increasing the atrial impulses rate makes the impulses to be closer so the probability to pass through slow pathway with the shorter refractory period will be higher. Also increasing γ will increase α. The term $\Delta\tau = \tau_1 - \tau_2$ is the difference between the two refractory periods. Defined the interarrival rate between the last blocked atrial impulse and the next one as t, there is a probability P for which $\tau_1 < t < \tau_2$ where the impulse could pass only from the slow pathway (since $t > \tau_1$), because the refractory period of the slow pathway is ended while the fast pathway is still refractory. According to Poisson property, the probability P for which the time t is included in the interval $\Delta\tau$ rises exponentially increasing τ_2 (as consequence $\Delta\tau$ raises):

$$P(t < \Delta\tau) = -e^{-\lambda\gamma\Delta\tau} \quad (3)$$

To complete the relationship between α and γ, the influence of the parameter $\Delta\tau_p$ was considered, leading to:

$$\alpha = 1 - \left[(1-\gamma)e^{-\lambda\gamma\left(\Delta\tau + \frac{\Delta\tau_p}{2}\right)}\right] \quad (4)$$

Since the estimated parameter from real data is α the inversion of the law is required.

The Lambert Function is defined to be the multivalued inverse of the function[7]:

$$y = f(W) = W(x)expW(x) \quad (5)$$

In case of real x, for $\frac{1}{e} \leq x < 0$ there are two possible real values of $W(x)$, see Fig. 1.2. The branch satisfying $-1 \leq W(x)$ is called the *principal branch* ($W_0(x)$) and the branch satisfying $W(x) \leq -1$ by $W_{-1}(x)$. Assuming:

$$z = 1 - \gamma; \Delta T = \Delta\tau + \frac{\Delta\tau_p}{2}$$

the equation will be:

$$\alpha = 1 - z\exp(-\lambda(1-z)\Delta T) \quad (6)$$

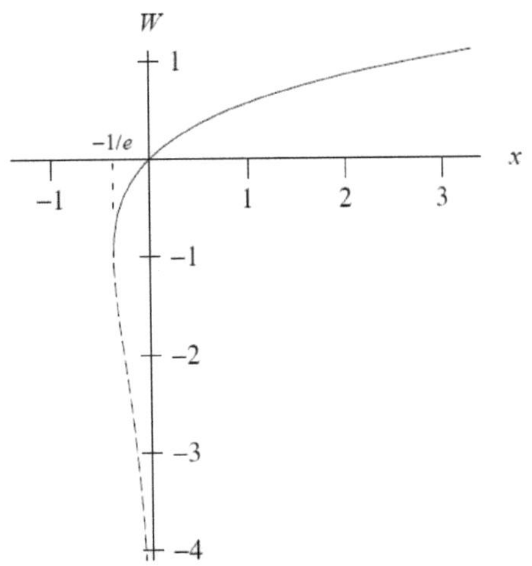

Figure 2: The two real branches of $W(x)$. The first dashed line is $W_0(x)$;the second continuous line is $W_{-1}(x)$[7]

Dividing the exponential:

$$\alpha' = z\exp(\lambda z \Delta T)$$

where $\alpha' = (1-\alpha)\exp(\lambda\Delta T)$. Changing the base $b = \exp(lambda\Delta T)$ it is possible to arrive to the compact form similar to the (1.2):

$$\alpha' = zb^z \tag{7}$$

Using the Lambert function property described in the article [7]:

$$x = W(x)b^{W(x)}; W(x) = \frac{W(x\log(b))}{\log(b)} \tag{8}$$

the final equation is:

$$\gamma = 1 - \frac{W((1-\alpha)\lambda\Delta T\exp(\lambda\Delta T))}{\lambda\Delta T} \tag{9}$$

C. Simulated data

Different parameter settings were tested, as shown in Table 1. For each parameter setting, a 30-min RR interval series was simulated. The parameter λ was incremented with 0.5 Hz step, the step of $\Delta\tau$ and $\Delta\tau_p$ was 0.01 s, and the probability γ varied with 0.1 step.

The parameter γ was given as input and the parameter α was estimated using Eq.4.

III. RESULTS

A. Simulation results

Figures 3 (a) and (b) show the comparison between the probability α obtained using Eq. 2 and that obtained in the

Table 1: Set of parameters used to alpha simulation

Inputs	Intervals	
λ	$5-10$	[Hz]
$\Delta\tau$	$0-0.5$	[s]
$\Delta\tau_p$	$0-0.2$	[s]
γ	$0-1$	

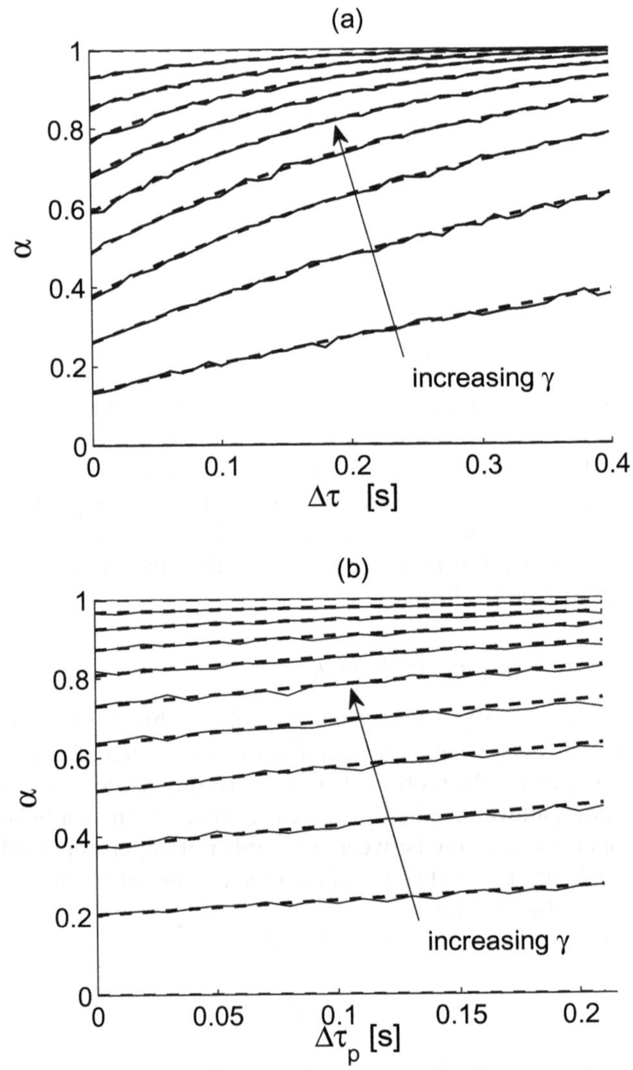

Figure 3: Trends of the theoretical (dashed lines) and real (solid lines) values of α as a function of the difference between (a) the two refractory periods ($\Delta\tau$) and (b) the prolongation of the two refractory periods ($\Delta\tau_p$). Each curve corresponds to a different value of γ that was varied from 0.1 to 0.9.

Table 2: Absolute error between the α obtained using Eq. 4 and the real value.

γ	*Error*
0.0	0.0029 ± 0.0005
0.1	0.0033 ± 0.0005
0.2	0.0031 ± 0.0006
0.3	0.0033 ± 0.0007
0.4	0.0035 ± 0.0007
0.5	0.0047 ± 0.0008
0.6	0.0038 ± 0.0008
0.7	0.0038 ± 0.0009
0.8	0.0045 ± 0.0010
0.9	0.0045 ± 0.0011

simulations, as a function of $\Delta\tau$ and $\Delta\tau_p$, respectively. It can be noted that the theoretical trend of α is very similar to the real one. Moreover, Table 2 shows the absolute error computed between the real α and its theoretical value, averaged on all simulations. These results confirm the goodness of the derived theoretical expression of α.

IV. DISCUSSION AND CONCLUSIONS

In this study, for the first time, we present a relationship between the probability γ of an atrial impulse to arrive at the slow pathway with the probability α (used in the model) of an impulse to pass through that pathway. A theoretical equation has been derived, showing that atrial and AV nodal electrophysiological properties play an important role. In fact, the derived relationship depends on both refractory periods and their prolongation and on the arrival rate of impulses to the AV node.

Further investigation will include application on real data.

REFERENCES

1. A. Rashidi I. Khodarahmi. Nonlinear modeling of the atrioventricular node physiology in atrial fibrillation *J. Theor. Biol.*. 2005;vol 232:545-549.
2. J. Lian V. Lang. Computer modeling of ventricular rhythm during atrial fibrillation and ventricular pacing *IEEE Transactions on Biomedical Engineering.* 1993;vol 53:1512–1520.
3. J. Lian D. Müssig. Heart rhythm and cardiac pacing: An integrated dual-chamber heart anc pacer model *Ann. Biomed. Eng.*. 2009;vol 37:64-81.
4. Corino V.D.A., Sandberg F., Mainardi L.T., Sörmo L.. An atrioventricular node model for analysis of ventricular response during atrial fibrillation *IEEE Trans. Biomed. Eng.*. 2011;vol 58:3386-3395.
5. Valentina D.A. Corino Luca T. Mainardi, Sörmo Leif. Atrioventricular nodal function during atrial fibrillation: Model building and robust estimation *Biomedical Signal Processing and Control.* 2013;8:1017-1025.
6. Sandberg F., Stridh M., Sörnmo L. Frequency Tracking of Atrial Fibrillation Using Hidden Markov Models *IEEE Trans. Biomed. Eng.*. 2008;55:502-511.
7. Corless R.M., Gonnet G.H., Hare D.E.G., Jeffrey D.J., Knuth D.E.. On the Lambert W function *Advances in Computational Mathematics.* 1996;5:329-359.

Evaluation of 3D Time-Reversal Focusing Method in Microwave Hyperthermia Treatment: Head and Neck Tumors

P. Takook, H. Trefna, and M. Persson

Chalmers, Göteborg, Sweden

Abstract— **Instructions for preparing papers for IFMBE-Proceedings Series are presented. They are intended to guide the authors in preparing the electronic version of their paper. Only papers prepared according to these instructions will be published in the paper and online version of Proceedings. Provide an abstract of your paper no longer than 300 words.**

Keywords— **Enter up to five keywords and separate them by commas.**

I. INTRODUCTION

In this paper we present time reversal focusing technique for hyperthermia treatment of deep seated tumors in 3D. This method which presented for 2D case has advantages of being fast and applicable to both continuous and pulsed waves. The quantitative evaluation of performance indicators in hyperthermia treatment is performed on a 3D model of head and neck, containing a tumor in tongue.

II. METHOD

We consider the 3D model for the head and neck. The antenna are arranged in two rings, with 8 antennas per ring, and modeled as both point sources and thin wire dipoles. To excite the virtual source, both sinusoidal and Gaussian pulsed wave are considered in the frequency range of 0.434-1 GHz. The size of the computational domain is 300*253*364 and the FDTD cells have a resolution of 1mm. Also we use the average power absorption (aPA) ratio to quantify the relative amount of energy absorbed in the tumor and the remaining tissue maximum index (RTMi) for the highest level of PA distribution in the remaining tissue respect to the tumor.

III. RESULTS

The effect of frequency and bandwidth is investigated for focusing in the center of the tumor. Between the center frequencies of 500 and 800 MHz, the best focusing in terms of high aPa value and low RTMi, is achieved for lower frequency of 500 MHz and with pulsed wave excitation of 300 MHz bandwidth. The thin wire dipole antenna under the same frequency and bandwidth gives better performance indicator values than the point source antennas.

IV. DISCUSSION

The highest aPa and smallest RTMi achieved using a pulsed wave excitation at center frequency of 500 MHz with BW of 300 MHz for focusing in the center of a tongue tumor with relatively large size.

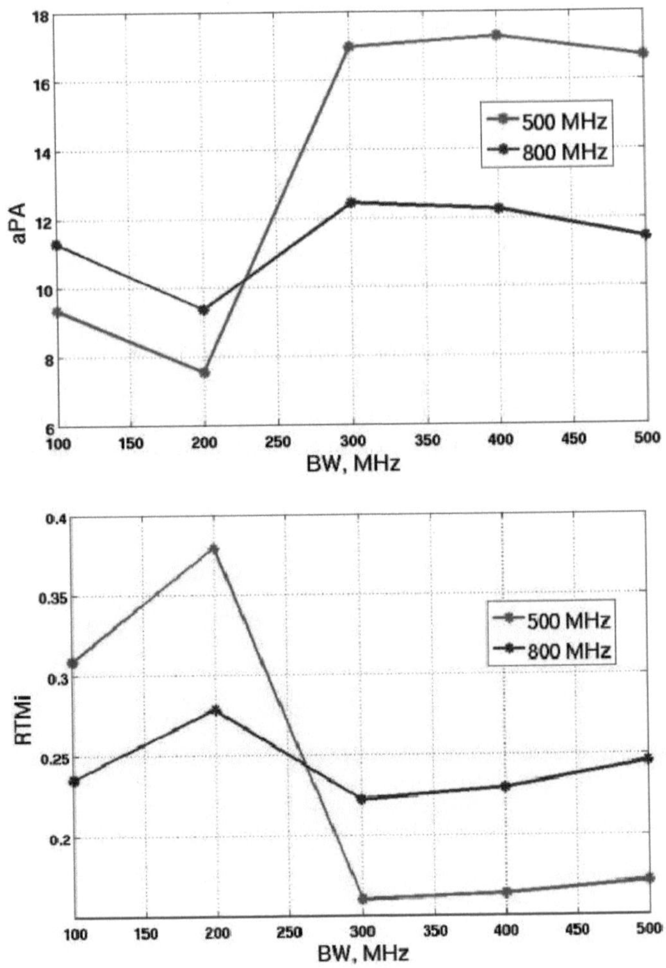

Fig. 1 Apa and RTMi computed at center frequencies of 500 and 800 MHz.

REFERENCE

[1] H. D. Trefna, J. Vrba, M. Persson, "Time-reversal focusing in microwave hyperthermia for deep-seated

H. Mindedal and M. Persson (eds.), *16th Nordic-Baltic Conference on Biomedical Engineering*,
IFMBE Proceedings 48, DOI: 10.1007/978-3-319-12967-9_28

Heart Rate Analysis by Sparse Representation for Acute Pain Detection

S. Tejman-Yarden[1,3], O. Levi[2], A. Beizerov[3], Y. Parmet[2], T. Nguyen[4], M. Saunders[5], Z. Rudich[6], J.C. Perry[1], D.G. Baker[7,8], and T. Moeller-Bertram[9]

[1] Department of Pediatrics, Division of Pediatric Cardiology, UC, San Diego, Rady Children's Hospital San Diego, CA
[2] Department of Industrial Engineering and Management, Ben Gurion University of the Negev, Beer Sheva, Israel
[3] Department of Biomedical Engineering, Ben Gurion University of the Negev, Beer Sheva, Israel
[4] Department of Electrical Engineering, University of California, San Diego, La Jolla, CA
[5] Department of Management Science Engineering, Stanford University, Stanford, CA
[6] Department of Anesthesiology and Critical Care, Soroka University Medical Center, Beer Sheva, Israel
[7] Department of Psychiatry, University of California, San Diego, La Jolla, CA
[8] Veterans Affairs Center of Excellence for Stress and Mental Health, San Diego, CA
[9] Department of Anesthesiology, University of California San Diego, La Jolla, CA

Abstract— **Objective pain assessment has not yet been achieved. Advanced signal processing methodologies, including the wavelet transform (WT) and the Orthogonal Matching Pursuit algorithm (OMP) were developed in the past two decades. The aim of this study was to apply and compare these time specific methods to heart rate samples of healthy subjects to investigate indicators of acute pain.** *Methods:* **15 adult volunteers participated in a study conducted in the pain clinic at a single center. Each subject's heart rate (HR) was sampled for 5 minutes baseline followed by a cold pressor test (CPT). Analysis was done by WT and OMP with a Fourier/Wavelet dictionary separately.** *Results:* **Data of 11 subjects was analyzed. Compared to baseline, WT analysis showed a significant wavelet coefficients frequency (WCF) increase during the pain incline period ($p<0.01$) and the entire CPT ($p<0.01$), with significantly higher amplitudes. OMP analysis showed a significant WCF increase during pain incline and decline periods ($p<0.01$, $p<0.05$) and the entire CPT ($p<0.001$), with suggestive higher amplitudes. Comparing the methods, during the baseline period there was a significant reduction of WCF using OMP analysis ($p<0.001$). Analysis by the two way ANOVA with repeated measures showed a significant proportional increase of WCF during the incline period and the entire CPT using OMP ($p<0.01$).** *Conclusion:* **HR analysis by both methods has successfully indicated the painful event at its onset, without delay. Statistical analysis proved OMP to be by far more specific. This is an initial study using OMP for pain detection. Further studies need to prove the validity of this system for pain detection in different settings.**

Keywords— **Pain, Heart Rate Variability, Wavelet Transform, Orthogonal Matching Pursuit.**

I. Introduction

Objective pain assessment has long been a challenge as pain is a subjective sensation depending on physical and mental factors. Physiological parameters such as skin conductance and heart rate (HR) have been studied but found unreliable [1]. Fourier transform of the HR and other spectral analysis methods (SAM) have not produced a reliable, real time pain detector [2,3], likely since biological signals are non-stationary signals, consisting of harmonic rhythms and sharp changes.

Unlike the SAM, the Wavelet transform (WT) can better identify acute events due to its properties of representing signals with a finite number of sudden short duration and time localized changes that differ from the baseline harmonics [4]. Novel methods using advanced Digital Signal Processing methodologies, such as the Orthogonal Matching Pursuit (OMP) algorithm [5,6], merge several different analysis methods to create a so-called over-complete dictionary, depicting the different elements of the signal in order to find an optimal representation. Thus, using both SAM (representing the harmonics of the signal) and WT (localizing the acute changes) may serve as a solution for acute pain onset detection.

Though not specific, the HR reacts sharply to acute painful stimuli. The aim of this study was to assess the HR of healthy young adults during pain from a cold pressor test, using WT and OMP. Our hypothesis was that the wavelet coefficients can be related to the acute painful stimulus and that analysis by OMP with an over-complete Fourier and Wavelet dictionary could improve pain detection. The two analysis methods, WT and OMP were applied separately to the HR as objective tools for acute, time related pain detectors and their performances were compared.

H. Mindedal and M. Persson (eds.), *16th Nordic-Baltic Conference on Biomedical Engineering*,
IFMBE Proceedings 48, DOI: 10.1007/978-3-319-12967-9_29

II. METHODS AND MATERIALS

A. Subjects

The study was approved by the ethics committee of the Soroka University Medical Center in Beer Sheva, Israel. The cold pressor test (CPT) was conducted in the Soroka University Medical Center Pain Clinic. Signal processing was performed later in the signal processing lab in the faculty of Biomedical Engineering in Ben Gurion University,Beer Sheva, in collaboration with the faculty of The Department of Electrical Engineering at the University of California San Diego. 15 healthy adult volunteers with no history or active medical disorders and no cardiac or neurologic history signed an informed consent before participation in the study. The participants' mean age was 28.1+/-3.5 years (10 males and 5 females).

B. Study Protocol and sampling system

During the study each subject was connected to an ECG polygraph using three skin surface electrodes and sampled at 1000 Hz using an 'Atlas Researches LTD' polygraph. The ECG samples were recorded in real-time though analysis was performed later in the signal processing lab. Each study was initiated with 5 minutes sampling in a quiet environment as baseline measurement. Then the subject was read the protocol & was requested to dip his or hers dominant hand into a bath of icy water. Pain was rated using the Visual Analog Score (VAS) [7]. Each subject performed the cold pressor test as long as he or she could endure it. Subjects were able to end the test or the sampling at any time.

C. Signal Processing

Each sample was reviewed before processing. In order for the results to be as clean & accurate as possible, short ECG segments that were noisy were cut out of the sample. Each subject's sample was then analyzed by the Matlab® program. From the entire ECG sample (baseline and protocol), by identifying the tip of the R wave, an R-R signal was generated, named tachogram. The tachogram of each subjects' sample was analyzed twice, by WT and by OMP. Each signal was reconstructed using up to 40 mixed coefficients with 1.5:1 to 1.8:1 ratio of wavelet preference. (The wavelet coefficients are multiplied by this ratio before being compared to the Fourier coefficients). Due to the nature of the WT, only the coefficients of the first 7 scales were used.

D. The Orthogonal Matching Pursuit Algorithm:

OMP generates an adaptive approximation of the signal using a greedy approach. It decomposes the signal by successive approximations into a linear expansion of waveforms that belong to a possibly redundant dictionary of functions (called atoms). These functions are selected to best represent the studied type of signal and its components. OMP is a greedy least-squares procedure that chooses the dictionary vectors one at a time, thus gradually reconstructing the signal using a set number of coefficients, each added gradually from the selected dictionaries (or transforms). Each coefficient is added layer by layer to build the reconstructed signal as close as possible to the original. After each selection step, the entire set of chosen coefficients is updated to make the signal approximation orthogonal to the residual vector. The residual norm is thereby minimized.

This study used OMP with a dictionary containing the Fourier and Wavelet transforms, as their combination gives the unique ability to discriminate between underlined harmonics and acute time-related events. The algorithm decides which specific, most correlative coefficient to utilize from the combined Fourier-Wavelet dictionaries at each iteration as it rebuilds the signal. The optimization problem that OMP solves approximately can be stated as

$$\min \|x\|_0 + \lambda \|r\|_2^2$$
$$s.t.: [A_1 \mid A_2]x + r = b \qquad (1)$$

where b is the given signal, A_1 and A_2 are matrices that correspond to the Fourier and Wavelet bases, x is the representation of b using the given over-complete dictionary, and r is the residual vector. $\|x\|_0$ is the number of non-zero entries in x, and λ is a sensitivity parameter. Since the size of b is n and x is $2n$ long, the linear system $[A_1 \mid A_2]x = b$ is under-determined and has multiple solutions. The solution of problem (1) finds a sparse representation of the signal b using the over-complete Fourier-Wavelet dictionaries and at the same time keeps the residuals small. The λ parameter penalizes the size of the residuals, so when λ is large it is expected that the optimal solution will have relatively small residuals, and when λ is small the residuals might be relatively large.

At step k=1 a single vector from the Fourier and Wavelet dictionary that has the highest correlation with b is chosen by solving the following problem:

$$\min_v \left(\min_c \|b - cv\|_2 \right) \qquad (2)$$

Where v is a vector of the matrix $[A_1 \mid A_2]$ and $c \in R$ (real number). Let $\hat{b}_1$ be the resulting best approximation of b at the first step, and let $e_1 = b - \hat{b}_1$ be the error after that first step. Similarly, let $\hat{b}_k$ be the resulting best approximation at the k-th step of b and let $e_k = b - \hat{b}_k$ be the associated error. At the k-th step the next vector is determined by solving

$$\min_{v}\left(\min_{c}\left\|e_{k-1} - cv\right\|_2\right) \quad (3)$$

As mentioned, $\hat{b}_k$ is the approximation of b using the k chosen vectors that correlate to the minimal e_k. The entire set of k chosen coefficients is then updated so that cv is orthogonal to e_k. The algorithm terminates when the approximation error norm is below a tolerance.

E. Data Analysis

This research studied the wavelet coefficients of both analysis methods. These were analyzed as markers for acute pain and were correlated to the personal VAS score reported by the subject. The two parameters that were correlated to the painful stimulus were the coefficients' occurrence frequencies (WCF) and coefficients' amplitudes.

Each subject's sample was divided into four periods. The first period was the baseline (VAS 0). The second period, "the pain incline period" included the anticipatory phase averaged 40 seconds during which the protocol was read to each subject and the initial phase of the cold pressor test (CPT) in which the subject reported incremental pain up to maximal pain (VAS 9-10). The third period was that of stable high pain levels. Finally, the last period was the pain decrement period, after the hand was removed out of the water and the VAS scores decreased as the pain resolved.

F. Statistical Analysis

Statistical analysis was performed to correlate the WCF and amplitudes with the painful stimulus. For each method, the WCF at the different time periods of the CPT was compared to the baseline period using the paired t-test. The ANOVA on the log scale was used to compare the performances of the two analysis methods in the different protocol periods. To analyze the amplitudes of the various stages the LMM (linear mixed models) was applied on the log scale of the amplitude level. This was used due to the unbalanced structure of the data with the need to take into consideration the random effect of the different participants. We used the log transforms of the different levels to achieve model assumptions.

III. Results

15 young adults participated in the study. The samples of 4 subjects were excluded, 2 subjects withdrew and 2 samples were excluded due to excessive motion artifacts. Thus, 11 subjects (9 males and 2 females) were included in the study analysis. For each subject a tachogram was created and wavelet coefficients were calculated using WT and OMP. The wavelet coefficients for each method were charted according to their occurrence time. Figure 1 shows typical results for of one of the subjects. The first diagram shows the wavelet coefficients calculated by WT and the second diagram shows the wavelet coefficients calculated by OMP. Each diagram charts the wavelet coefficients versus time. The study protocol and VAS charting are indicated on the bottom of the figure.

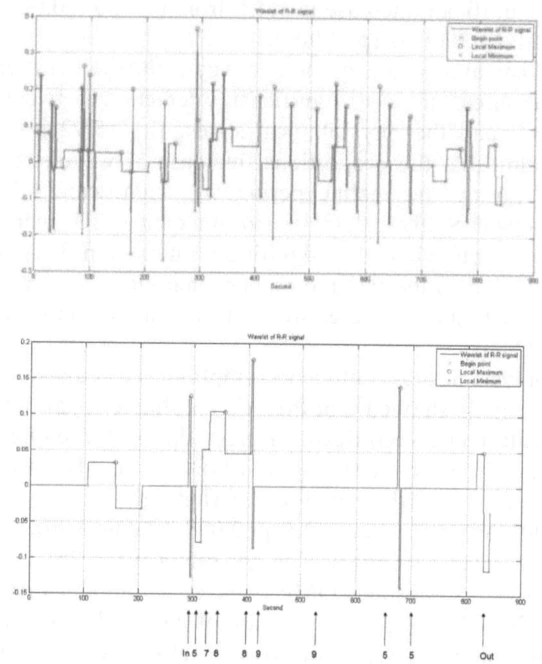

Fig. 1 Reconstruction of a tachogeam of a subject the WT (upper diagram) and the OMP (lower diagram). The protocol is indicated on the bottom.

Statistical analysis by a paired t-test of the WCF during the baseline period showed there was a significant reduction of WCF using OMP compared to WT ($p<0.001$). WCF was 0.0433 ± 0.054 Hz using WT and 0.0056 ± 0.0036 Hz using OMP.

Analysis of the entire signal by WT showed there was a significant increase in WCF during the pain incline period compared to the baseline period, ($p<0.01$). The baseline WCF was 0.0433 ± 0.054 Hz and the pain incline period WCF was 0.063 ± 0.054 Hz. There were no statistically significant changes regarding the stable maximal pain period compared to baseline and none regarding the pain decline period compared to baseline or to the stable maximal pain period. The WT showed a significant increase of WCF during the entire CPT compared to baseline, ($p<0.05$), from 0.043 ± 0.054 Hz to 0.059 ± 0.052 Hz during the entire CPT.

Analysis by OMP showed there was a significant increase in WCF during the pain incline period compared to baseline, ($p<0.01$). The baseline WCF was 0.0056 ± 0.0036 Hz and the pain incline period WCF was 0.055 ± 0.068 Hz. WCF dropped during the stable persistent maximal pain period (though not statistically significantly) and significantly increased again during the pain decline period to 0.024 ± 0.017 Hz, ($p<0.05$) marking out the periods of acute physiological changes. Comparison of WCF during the entire CPT to baseline using a paired t-test showed there was a significant increase in WCF from 0.0056 ± 0.0036 Hz to 0.031 ± 0.039 Hz ($p<0.001$).

Statistical analysis by the two way ANOVA with repeated measures, showed significant interaction between the methods and the physiological states, $F_{1,10}=5.92$ $p=0.035$, indicating that the proportional increase of wavelet coefficients during the incline period was significantly higher using OMP. Statistical analysis of the entire CPT, using this method again yielded a significant interaction $F_{1,10}=9.881$ $p<0.01$, indicating that the proportional increase of wavelet coefficients during the entire CPT was significantly higher using OMP.

Analysis of the coefficients' amplitudes calculated by the two methods showed that the WT amplitudes analysis by a log scale had a significant increase during the entire CPT compared to baseline ($F_{3,290}=6.413$ $p<0.001$), this with some preference to the incline period ($t_{290} = 3.908$ $p=0.001$(incline), $t_{290} = 2.6$ $p=0.039$ (stable pain), $t_{290} = 2.85$ $p=0.023$ (decline)). The net effect was 17% increase compared to baseline. OMP wavelet amplitude difference analysis had a suggestive increase of the amplitude between baseline and the CPT, the overall effect was marginally significant ($F_{3,106}=2.368$ $p=0.075$) this was probably subject to the small number of wavelet coefficients calculated by OMP during baseline.

IV. Discussion

This study focused on HR analysis for acute pain detection using WT and OMP integrating the Wavelet and the Fourier transforms. In both methods we examined the time-related wavelet coefficients to examine whether real time detection of a painful stimulus could be made. Both analysis methods showed a significant increase of WCF during the initial pain incline period and during the entire cold pressor test, as compared to the baseline period. Moreover a significant rise in the wavelet coefficients' amplitudes was also observed, mostly when using the WT.

When comparing the two methods, a striking finding correlated with our hypothesis: during the baseline period there was an absolute reduction of wavelet coefficients number using OMP analysis as compared to WT analysis. This was attributed to the fact that during the baseline period there were minimal acute events and thus the wavelet coefficients charted by WT were represented now by OMP's Fourier transform coefficients, as functional harmonics.

Further comparison of the two methods by the two-way ANOVA with repeated measures showed a significant proportional increase of WCF during the incline period using the OMP algorithm as compared to the WT analysis. When the same analysis, was applied to the data of the entire CPT versus the baseline, this again yielded a significant proportional increase using OMP. These findings demonstrate the advantage the OMP analysis has over the basic WT in analyzing a biological signal and decomposing it to its basic components – harmonics and acute events.

Limitations: We should emphasize though that by using the tachogram this system analyzed only the HR. Any other information presented in the entire ECG signal like AV conductance, QRS morphology & amplitudes, or ST segment changes was eliminated. This was a group of healthy young adults, thus different settings should be tested to evaluate the HR response in other groups. Moreover the HR signal is highly nonspecific and may react differently to other stimuli and stress triggers.

In conclusion HR analysis by both methods was able to clearly identify the painful event at its onset, without delay. This is the first study in which the OMP analysis method was used for the acute detection of painful stimuli and proved it to be by far more sensitive, bringing us one step further in the quest of an objective acute pain detector.

References

1. Tousignant-Laflamme Y, Rainville P, and Marchand S. Establishing a link between heart rate and pain in healthy subjects: a gender effect. J Pain 2005, 6:341–347.
2. Ray G, Das G, Ray P. Design of ECG based anesthesia monitor/ pain monitor. Conf Proc IEEE Eng Med Biol Soc. 2004;1:25-8.
3. Storella RJ, Shi Y, O'Connor MD, Pharo GH, Abrams JT and Levitt J. Relief of chronic pain may be accompanied by an increase in a measure of heart rate variability, Anesth Analg 1999;89:448–50
4. Magosso, E, Ursino, M, Zaniboni, A, and Gardella, E. A wavelet-based energetic approach for the analysis of biomedical signals: Application to the electroencephalogram and electrooculogram. Applied Mathematics and Computation. 2009;207(1):42–62
5. Mallat SG and Zhang Z. Matching Pursuits with time-frequency dictionaries, IEEE Transactions on Signal Processing, December 1993; 3397-3415.
6. Chen SS, Donoho DL, and Saunders MA. Atomic decomposition by Basis Pursuit, SIAM J. Sci. Comput. 1988;20(1):33-61
7. Davis P and Walsh D. Cancer pain: How to measure the fifth vital sign. Cleveland Clinic. J of Med 2004; 71(8): 625-632.

Optical Urea Rebound Estimation during Dialysis

R. Tomson[1], F. Uhlin[2], and I. Fridolin[1]

[1] Department of Biomedical Engineering, Tallinn University of Technology, Tallinn, Estonia
[2] Department of Medicine and Health Sciences, Linköping University,
Department of Nephrology UHL, Linköping, Sweden

Abstract— **The aim of the study was to explore the connection of urea rebound and the difference between spKt/V and eKt/V and also the possibility of utilizing UV-absorbance measurements to assess urea rebound. Ten patients of chronic three-times-a-week hemodialysis (HD) were studied. On-line UV-absorbance of spent dialysate was monitored. Single-pool Kt/V (spKt/V), equilibrated Kt/V (eKt/V) and the percentage difference between spKt/V and eKt/V (ΔKt/V) were calculated. Urea rebound was calculated based on urea concentration in blood (R_b) and UV-absorbance in spent dialysate (R_a). ΔKt/V and R_b were not statistically different. Also, R_a and R_b were not statistically different. In summary, the results show that it is possible to assess post-dialysis urea rebound in blood based on UV-absorbance in spent dialysate, which may offer the opportunity to estimate the true dialysis dose and a more personalized approach to the dialysis treatment.**

Keywords— **hemodialysis monitoring, rebound, Kt/V, urea, UV-absorption**

I. Introduction

Urea, a low-molecular weight metabolic end product of the catabolism of proteins, is considered to be the most suitable marker for uremic toxins in the range of low-molecular weight solutes [1]. Urea Kt/V is viewed as a sensitive measure of the overall dialysis dose that characterizes dialysis adequacy [2]. Traditionally, Kt/V is derived from formal urea kinetic modeling (UKM), which is based on blood samples at the start and end of dialysis [1].

If the immediate post-dialysis urea concentrations are used for the calculation of dialysis dose, it can be significantly overestimated because of the increase in blood urea concentration – urea rebound – which occurs after completion of the HD session. Urea rebound is complete within 30-60 min after the cessation of HD, which means that the most accurate way for the calculation of Kt/V would be to wait up to 60 minutes after the completion of HD before drawing the post-dialysis sample. However, this approach is impractical for patients and dialysis facilities.

In order to avoid the delay of waiting for an equilibrated post-dialysis blood sample, algorithms for anticipating post-dialysis rebound of urea have been developed [3, 4]. As the percentage value of rebound relative to the fall in urea concentration during HD approximates the percentage difference between single-single pool Kt/V (spKt/V) and equilibrated Kt/V (eKt/V) [5], this information could be used to estimate the true dialysis dose.

The Smye algorithm [3] estimates the post-dialysis equilibrated urea concentration in blood based on conventional pre- and post-dialysis blood samples and an additional intradialytic blood sample. The drawback of the Smye algorithm is that it suffers from the effects of small urea concentration measurement errors [6]. The Smye algorithm has also been modified for the use together with a continuous urea sensor [7] showing good agreement between the estimated equilibrated urea concentration and urea concentration 25-40 min following termination of dialysis.

There is a need for an instrument capable of directly and easily assessing post-dialysis urea rebound without the need to have the patient wait 30-60 min after the treatment and without repeated blood samples. An optical method utilizing UV-absorbance has been proposed for the monitoring of dialysis adequacy [8, 9]. A good linear relationship has been found between UV-absorbance and dialysate urea concentration in the wavelength range 210-330 nm, with the highest correlation at 280-320 nm [10]. It has been shown that due to the good correlation between UV-absorbance and urea concentration in dialysate, the latter can be estimated from UV-absorbance measurements even if the UV-technique does not measure urea itself [11]. Moreover, urea concentration in spent dialysate is a fixed fraction of arterial urea concentration as long as dialysate flow rate, dialyser clearance and recirculation rate remain unchanged [7].

This study was undertaken to explore the connection of urea rebound and the difference between spKt/V and eKt/V and also the possibility of utilizing UV-absorbance measurements to assess urea rebound.

II. Subjects and Methods

A. Subjects

Ten patients, four females and six males, mean age 60±19 years, on chronic three-times-a-week HD were studied at the Department of Nephrology, University Hospital, Linköping, during a total of 30 sessions. Six patients, i.e. 18 treatments, were dialysed by a low-flux membrane (Polyflux, 17L, Gambro, Sweden) with an effective membrane

H. Mindedal and M. Persson (eds.), *16th Nordic-Baltic Conference on Biomedical Engineering*,
IFMBE Proceedings 48, DOI: 10.1007/978-3-319-12967-9_30

area of 1.7 m^2 and four patients, i.e. 12 treatments, with high-flux membranes (Nephral 300, HOSPAL Industrie, Meyzieu, France and Tricea 150, Baxter Health Care Corp., IL, USA) with an effective membrane area of 1.3 m^2 and 1.5 m^2, respectively. Treatment durations ranged from 240 to 270 min. The dialysate flow was fixed at 500 ml/min and effective blood flow varied between 200 to 350 ml/min. The type of dialysis machine used was Fresenius 4008H (Fresenius Medical Care, Germany).

The Regional Ethics Committee, Linköping, Sweden approved the study protocol and informed consent was obtained from all patients.

B. Sampling and laboratory analysis

Samples of blood were taken at before the start of HD, at the end of HD and 30 min after the end of HD. The blood samples were sent to the laboratory for analysis within 2-4 h. Laboratory's standard sampling procedures were followed. The concentration of urea was determined at the Clinical Chemistry Laboratory at the Linköping University Hospital using a standardized method. The accuracy of the method for the determination of urea in blood was ±5 %.

C. UV-absorbance monitoring

The UV-instrumentation has been described earlier [12]. The wavelength 297 nm was used. The sampling frequency was set at two samples per minute.

The baseline was measured a few minutes before the start of each dialysis treatment on the flowing pure dialysate (reference solution) when the temperature and conductivity had been stabilized and the sodium and bicarbonate level had been preset according to the patient records.

The obtained UV-absorbance values were processed and presented on computer screen by a PC incorporated into the spectrophotometer using Kontron software (UVIKON 943, version 7.0 for Windows; Kontron Instruments, Italy). Data were then transformed to an Excel file at the end of the treatment.

D. Data analysis

Some of the measured values (absorbance or concentration) were excluded from data before analysis. The exclusion criteria were incorrect or illogical values of the measured concentration or absorption, e.g. sampling coexisting with self-tests of the dialysis machine.

Single-pool Kt/V (spKt/V) was calculated according to [1] as

$$spKt/V = -\ln\left(\frac{C_t}{C_0} - 0.008T\right) + \left(4 - 3.5\frac{C_t}{C_0}\right)\frac{\Delta BW}{BW} \quad (1)$$

where C_0 and C_t are blood urea concentrations before and at the end of the dialysis, respectively, measured in mmol/l, T is treatment time in hours, dBW intradialytic weight loss in kilograms and BW end-session body weight in kilograms.

Equilibrated Kt/V (eKt/V) was calculated according to [1] as

$$eKt/V = spKt/V - \left(0.6\frac{spKt/V}{T}\right) + 0.03 \quad (2)$$

The difference between spKt/V and eKt/V (ΔKt/V) was expressed as

$$\Delta Kt/V = \frac{spKt/V - eKt/V}{eKt/V}100\% \quad (3)$$

Urea rebound (R) was expressed relative to C_t as

$$R = \frac{C_{eq} - C_t}{C_t}100\% \quad (4)$$

where C_{eq} is the equilibrium concentration of urea at the end of rebound phase. Rebound was calculated based on urea concentration in blood samples (R_b) and UV-absorbance in spent dialysate (R_a). In case of R_a urea concentrations were substituted by UV-absorbance values.

In order to estimate urea rebound based on UV-absorbance in spent dialysate a substitute value for C_{eq} (A_{eq}) was calculated according to the Smye algorithm [3] where urea concentrations were substituted by UV-absorbance values

$$A_{eq} = A_0 e^{-\lambda t} \quad (5)$$

so that A_0 is the average value of 2 to 6 min from the beginning of HD and t is the duration of HD in minutes. λ was obtained by line fitting based on on-line UV-signal from 60 min to the end of HD session.

Student's t-test for dependent samples was used to compare means for estimated parameters and $p<0.05$ was considered significant. Individual differences in ΔKt/V and R_a compared to R_b were also examined using Bland and Altman analysis [13].

For the analysis Excel (version 2003 for Windows) was used.

III. Results

Average spKt/V was 1.45±0.23 and average eKt/V was 1.27±0.20. Average ΔKt/V was 13.76±1.12 % and it was not statistically different from R_b (p=0.57), which was 13.25±4.95 %. Figure 1 shows the Bland-Altman plot of the differences between R_b and ΔKt/V. The mean difference between R_b and ΔKt/V was -0.57±4.38 %.

Average R_a was 13.20±7.54 % and it was not statistically different from R_b (p=0.79). Figure 2 shows the Bland-Altman plot of the differences between R_b and R_a. The mean difference between R_b and R_a was -0.43±8.15 %.

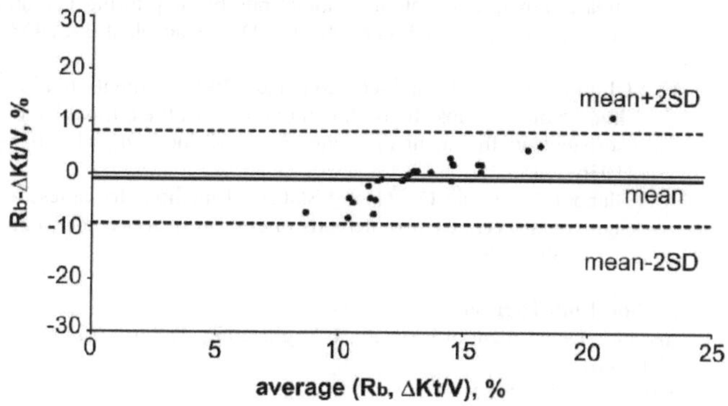

Fig. 1 Bland-Altman plot of the differences between R_b and ΔKt/V (N=26)

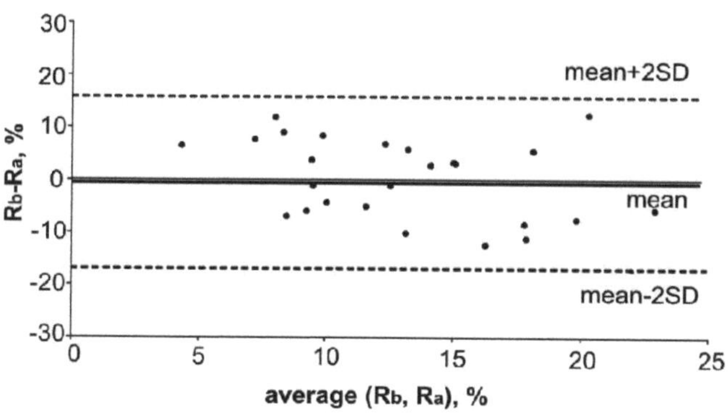

Fig. 2 Bland-Altman plot of the differences between R_b and R_a (N=26)

IV. Discussion

The present study investigated the connection of urea rebound and the difference between spKt/V and eKt/V and also the possibility of utilizing UV-absorbance measurements to assess urea rebound.

The results indicated that: (i) urea rebound percentage value approximates the percentage difference between spKt/V and eKt/V; (ii) it is possible to assess post-dialysis urea rebound in blood based on UV-absorbance measurements in spent dialysate.

It has been suggested previously that the percentage value of urea rebound approximates the percentage difference between spKt/V and eKt/V [5]. The results of this study support this assumption, as ΔKt/V was not statistically different from R_b (p=0.57).

It has been show previously that it is possible to monitor urea concentration in spent dialysate with the UV-absorbance technique [11]. The optical method for monitoring dialysis adequacy [8, 9] offers the possibility to continuously follow the urea elimination profile without the need for disposables of chemicals. As urea concentration in spent dialysate is a fixed fraction of arterial urea concentration as long as dialysate flow rate, dialyser clearance and recirculation rate remain unchanged [7] it is also feasible to estimate urea rebound in blood based on UV-absorbance in spent dialysate. This assumption is supported by the results, as R_a was not statistically different from R_b (p=0.79). Thus, the results of the present study indicate the possibility of assessing urea rebound based on UV-absorbance measurements in spent dialysate.

As the results of this study show that R_b approximates ΔKt/V and the possibility of assessing urea rebound based on UV-absorbance measurements in spent dialysate exist, it can be suggested R_a could be utilized to estimate ΔKt/V. This information could be used to estimate the true dialysis dose and it would enable a more personalized approach to the dialysis treatment.

V. Conclusion

The results suggest that it may be feasible to assess post-dialysis urea rebound in blood based UV-absorbance measurements in spent dialysate, which may offer the opportunity to estimate the true dialysis dose and a more personalized approach to the dialysis treatment. To validate the results using a larger database will be an issue of further studies. The merits of the described method are that it does not need blood samples or the patient to wait 30-60 minutes after the completion of HD before the drawing the post-dialysis sample.

Acknowledgment

The study was partly supported by the Estonian Science Foundation Grant No 8621, by the European Union through the European Regional Development Fund and institutional research funding IUT19-02. The study was partly supported by the County Council of Östergötland, Sweden. The authors have no conflicts of interest to report.

Conflict of Interest

The authors declare that they have no conflict of interest.

References

1. European Best Practice Guidelines of Haemodialysis, ERA-EDTA. (2002) II.1 Haemodialysis dose quantification: small solutes. Nephrol Dial Transplant 17: 16-31
2. NKF KDOQI guidelines. Clinical practice guidelines for hemodialysis adequacy, update 2006. Guideline 3. Methods for postdialysis blood sampling at http://www.kidney.org
3. Smye SW, Evans JH, Will E et al. (1992) Paediatric haemodialysis: estimation of treatment efficiency in the presence of urea rebound. Clin Phys Physiol Meas 13: 51-62
4. Daugirdas JT, Schneditz D. (1995) Overestimation of hemodialysis dose depends on dialysis efficiency by regional blood flow but not by conventional two pool urea kinetic analysis. ASAIO J 41: M719-M724
5. Alloatti, S, Molino A, Manes M et al. (1998) Urea rebound and effectively delivered dialysis dose. Nephrol Dial Transplant 13: 25-30
6. Gotch FA, Keen ML (2005) Kinetic modeling in hemodialysis, in Nissenson AR, Fine RN (ed), Clinical dialysis, New York, McGraw-Hill, pp. 153-202.
7. Garred LJ, Canaud B, Bosc JY et al. (1997) Urea rebound and delivered Kt/V determination with a continuous urea sensor. Nephrol Dial Transplant 12: 535-542
8. Fridolin I, Magnusson M, Lindberg L-G (2002) On-line monitoring of solutes in dialysate using absorption of ultraviolet radiation: technique description. Int J Artif Organs 25:748-761
9. Castellarnau A, Werner M, Günthner R et al (2010) Real-time Kt/V determination by ultraviolet absorbance in spent dialysate: technique validation. Kidney Int 78: 920-925
10. Fridolin I, Lindberg L-G (2003) On-line monitoring of solutes in dialysate using wavelength-dependent absorption of ultraviolet radiation. Med Biol Eng Comput 41: 263-270
11. Uhlin F, Fridolin I, Lindberg L-G et al. (2005) Estimating total urea removal and protein catabolic rate by monitoring UV absorbance in spent dialysate. Nephrol Dial Transplant 20: 2458-2464
12. Uhlin F, Fridolin I, Lindberg L-G et al. (2003) Estimation of delivered dialysis dose by on-line monitoring of the ultraviolet absorbance in the spent dialysate. Am J Kidney Dis 41: 1026-1036
13. Bland JM, Altman DG (1986) Statistical methods for assessing agreement between two methods of clinical measurement. Lancet 1: 307-310

Author: Ruth Tomson
Institute: Department of Biomedical Engineering, Tallinn University of Technology
Street: Ehitajate tee 5
City: 19086 Tallinn
Country: Estonia
Email: ruth@cb.ttu.ee

Factors Affecting the Statistical Analysis of Ultrasonic Backscattering Signals and Imaging

Ya-Ting Tsai, Jiue-Jia Wu, Yi-Hsun Lin, and Shyh-Hau Wang

Department of Computer Science and Information Engineering & Institute of Medical Informatics, National Cheng Kung, University, Tainan, Taiwan

Abstract—**Statistical distribution of ultrasonic backscattering signals has been demonstrated capable of characterizing variations of density and arrangement of scatterers in biological tissues. The statistical analysis of ultrasound signals has also found with less dependency on the attenuation effect. Yet, as the employed ultrasound frequency and pulse duration were increased, several factors could further affect the precise estimation of the statistical parameters. To further investigate the addressed issues, experiments were arranged and performed from tissue-mimicking phantoms and porcine livers. Various duty cycles, including 1, 3, 5, and 10%, of tone bursts at 1 KHz pulse repetition frequency corresponding to ultrasound frequencies of 3.5, 7.5, and 10 MHz were adjusted for driving the transducers. The tissue-mimicking phantoms were fabricated, which consisted of gelatin and glass beads of 16 and 64 scatterers/mm^3 and those of porcine livers with either healthy or pathological fibrosis were obtained from local slaughter house. Various thickness of Silicone plates with the attenuation coefficient of 1.62 dB/mm•MHz were placed on the surface of objects to be measured. Nakagami statistical model, including shape parameter (Nakagami-*m*) and parametric imaging, was implemented to assess variations of the probability density function (PDF) estimated from the acquired ultrasonic backscattering signals. Results of phantoms indicated that the attenuation could significantly vary the shape of PDF of backscattered envelopes. Especially, large attenuation effect was found corresponding to those broader incident ultrasound bandwidths excited by the monocycle signal; whereas the effect is substantially reduced as the tone bursts were more than 3 cycles. Results of porcine livers indicated that the Nakagami-*m* increased with the increasing ultrasound frequencies and bandwidth, and that those associated PDFs were nearly pre-Rayleigh distributed. Both phantoms and porcine livers results consistently demonstrated that the use of 3 cycles tone bursts for exciting transducers may achieve the most appropriate performance to accommodate a tradeoff between attenuation effect and image resolution. Current study also verified that the operational modes of incident ultrasound need to be properly assured before that the statistical model may be further applied to clinical applications.**

Keywords—**Ultrasound backscattering, Statistical model, Probability distribution function, parametric imaging.**

I. INTRODUCTION

The ultrasound signals backscattered from tissues usually exhibited as a form of undesired random modulation, known as speckle [1, 2]. Speckle typically is a form of granular texture in ultrasonic B-mode image, which is generally explained as a result of scatterings interference phenomenon. On the other hand, ultrasonic backscattered signals are well known to correlate to tissue structures and their scattering properties, and they can be expressed as the sum of signals from each single scatterer in the resolution cell [3, 4]. Therefore, tissue properties associated with normal, benign, and malignant states may be characterized quantitatively by calculation of acoustic parameters and speckles of the acquired backscattered signals [5].

In addition, a statistical distribution can be used to model the probability density function (PDF) of ultrasonic backscattered envelopes, which correlate with the properties, concentration, and distribution of the scatterers in biological tissues [6]. Among various statistical models, Nakagami statistics [4] is a more general model with less dependency on the attenuation effect. The Nakagami statistical model has been utilized to characterize tissues, such as breast [7], blood [8], and wound tissue [9]. A shape function, namely Nalagami parameter (Nakagami-*m*), has also found able to be utilized to form parametric imaging for quantitatively detailing the local distribution of tissues [10]. Despite of its usefulness, as the employed ultrasound frequency and pulse duration were increased, several factors could further affect the precise estimation of the statistical parameters. This is due to ultrasound attenuation tends not only to decrease the energy of the ultrasound, but also to increase the variance of the PDFs. Hence, the variance of Nakagami-*m* is conceptually to be increased. Furthermore, the broadband of ultrasound signals corresponds to the excitation of transducer by a pulse and that of narrowband is to those of driving signals with many cycles of sinusoidal wave. These two effects may affect the shape of statistical distribution and that of the corresponding Nakagami-*m*.

In the present study, factors, including attenuation, frequency, and bandwidth of incident waves, affecting the statistical analysis of ultrasound backscattered signals were extensively explored. Experiments were firstly performed from phantoms with various scatterer concentration. Different attenuation effects on the statistical parameter was investigated by placing the phantom with a silicone plate with a certain attenuation. The frequencies of employed transducers included 3.5, 7.5, and 10 MHz driven by various

H. Mindedal and M. Persson (eds.), *16th Nordic-Baltic Conference on Biomedical Engineering*,
IFMBE Proceedings 48, DOI: 10.1007/978-3-319-12967-9_31

cycles of sinusoidal waves. Further experiments were conducted to investigate scattering properties of healthy and fibrosis liver tissues.

II. MATERIALS AND METHODS

A. Phantom and Porcine Liver

Two phantoms added with concentrations of glass beads, 16 and 64 scatterers/mm^3, were prepared for experiments. The procedure of making phantoms can be found in previous work [6], in which that the phantom is mainly composed of gelatin powder, double distilled water, glass beads (mean diameter of 44 μm), and formalin.

The pathological fibrosis porcine liver tissues (n=5) were obtained from a local slaughter house and that of the healthy porcine liver tissues (n=5) were provided from a traditional local market. These tissue samples were examined by the veterinarian.

B. Data Acquisition

A schematic diagram detailing the experimental arrangement of the ultrasound system was shown in Fig.1. Experiments were performed respectively using 3.5, 7.5, and 10 MHz single-element focused transducers to generate and receive the ultrasonic signals. The characteristics of 3.5 and 7.5 MHz transducers are given in Table I. Signals generated and amplified respectively from the arbitrary function generator (AFG 3252, Tektronix, USA) and power amplifier (325 LA, Electronics & Innovation, USA) were utilized to drive the transducers. The electronic expander was applied to reduce noise, and that of the electronic limiter was used for protecting the A/D converter. The radio frequency (RF) signals were filtered by a band-pass filter and then digitized by an 8-bit analog-to-digital converter (PXI 5152, National Instruments, TX, USA) at 100 MHz sampling rate. The B-mode ultrasonic images from each sample scattering phantom and porcine liver were acquired by a swept scanning mode in which A-line signals of different locations were obtained from the transducer and positioner. In addition, the transducers were flexibly moved in two directions controlled by the two axes of servo motors and actuators. The program developed for both data acquisition and motor control were implemented using LabView software (National Instruments, TX, USA).

Various thickness of silicone plates with the attenuation coefficient of 0, 0.23, 0.91, 4.23, and 8.48 dB were placed on the surface of the phantom or porcine liver to be measured. Various sinusoidal cycles, including 1, 3, 5, and 10 cycles,

Table I. Characteristics of transducers

Frequency	3.5 MHz	7.5 MHz
Manufacturer	PANAMETRICS	PANAMETRICS
Model	V382	V320
Center frequency (MHz)	4.2	6.8
Aperture size	0.5”	0.5”
f#	1.7	1.5
Depth of focus (mm)	20.49	17.67
-6 dB bandwidth (MHz)	2.22-5.22	3.8-9.12

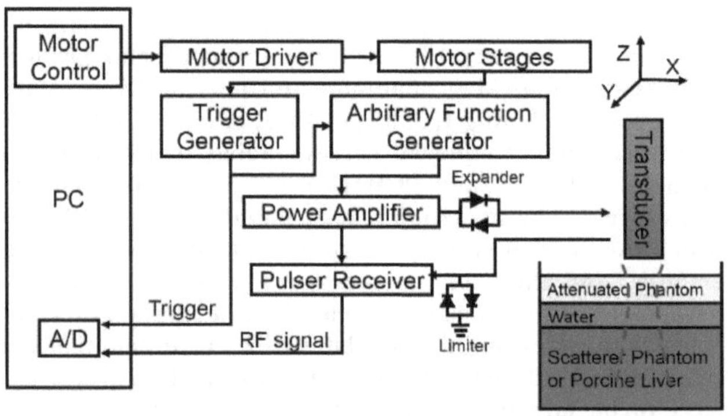

Fig.1, The shematic diagram of the experimental arrangement.

were generated by the arbitrary function generator to drive the transducer.

C. Nakagami Distribution

Statistical analysis, frequently adopted for the probability distribution function (PDF) of ultrasonic backscattered envelopes, has been generally utilized for tissue characterization. The tissue scatterers were investigated, and that of their influence on the frequency-dependent attenuation associated with the Nakagami distribution was analyzed. Nakagami distribution is given by (1)[4], where Γ and U denote respectively the gamma function and unit step function.; E denotes the statistical mean, R is the envelope of the backscattered signals, Ω is a scaling parameter (as given in (2)), and m is the Nakagami parameter (as given in (3)).

$$f(r,m,\Omega)=\frac{2m^m r^{2m-1}}{\Gamma(m)\Omega^m}\exp(-\frac{m}{\Omega}r^2)U(r) \tag{1}$$

$$\Omega=E(R^2) \tag{2}$$

$$m=\frac{[E(R^2)]^2}{E[R^2-E(R)^2]^2} \tag{3}$$

Nakagami parameter is a shape function, in which $m = 1$ relates the acquired signals to be Rayleigh distributed; that of m smaller than 1 correlates to pre-Rayleigh distribution, and the Nakagami parameter larger than 1 associates the backscattering statistics to post-Rayleigh distribution [7].

III. RESULTS

Results of phantoms and porcine livers indicated that the attenuation could significantly vary the shape of PDF of backscattered envelopes. As there are more than ten effective numbers of the scatterers in the resolution cell, the corresponding Nakgami-*m* tended to approach to 1 of Rayleigh distribution. Two kinds of scatterer phantoms with concentration of 16/mm^3 and 64/mm^3 were arranged, in which the corresponding resolution cells for the 3.5 MHz, 7.5MHz, 10MHz transducers were respectively 0.31 mm^3, 0.15 mm^3, 0.06mm^3, and those of the corresponding effective number of scatterer phantoms contained approximately 0.64 and 2.56 scatterers, respectively. The results of Nakagami-*m* for 16/ mm^3 and 64/mm^3 phantoms estimated from 3.5 MHz and 7.5 MHz transducers were shown in Fig. 2. The attenuation coefficient of 0, 0.23, 0.91, 4.23, 8.46 dB for the employed 3.5MHz transducer were measured from those ilicone plates of various thickness including 0, 0.5, 1, 2, 3 mm. The corresponding attenuation for 7.5 MHz ultrasound were measured to be 0, 0.35, 1.33, 4.75, 9.34 dB and those of for 10 MHz were 0, 0.36, 1.36, 5.74, 10.06 dB. The effect of attenuation on the variation of Nakagami-*m* may be found corresponding to the broadband ultrasound with the excitation of one cycle signal. The concentration effect also can be found, in which 16/ mm^3 phantoms revealed that the resolution cell of 7.5 MHz transducer tended to be pre-Rayleigh distributed.

Results of the subsequent pathological fibrosis porcine liver and healthy porcine liver tissues were given in Fig. 3. The Nakagami-*m* increased with the increase of excitation cycles of from 1 to 10 cycles. Similar tendency of results demonstrated that the higher the attenuation and that the higher the corresponding Nakagami-*m*. Higher attenuation contributed to larger Nakagami-*m* corresponds to those of broadband ultrasound.

In the study, Nakagami-*m* was estimated to quantify the properties of phantoms, *in vitro* porcine liver tissues. Nakagami-*m* of the ultrasonic envelopes acquired from the healthy porcine livers and the pathology porcine livers by three different frequencies have shown that the resultant values are higher for those of narrowband driven by ten cycles than those of by 1, 3, 5 cycles.

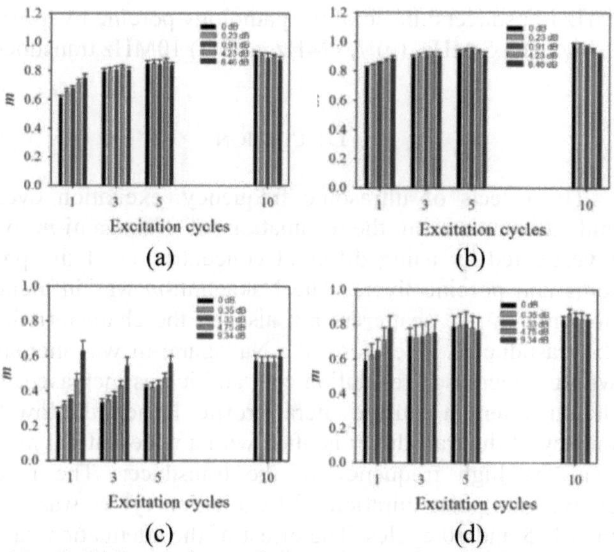

Fig.2 Nakagami-*m* of the ultrasonic envelopes was acquired from (a) 16/ mm^3 (b) 64/ mm^3 phantoms measured by 3.5MHz transducer; those of (c) and (d) 7.5MHz transducer.

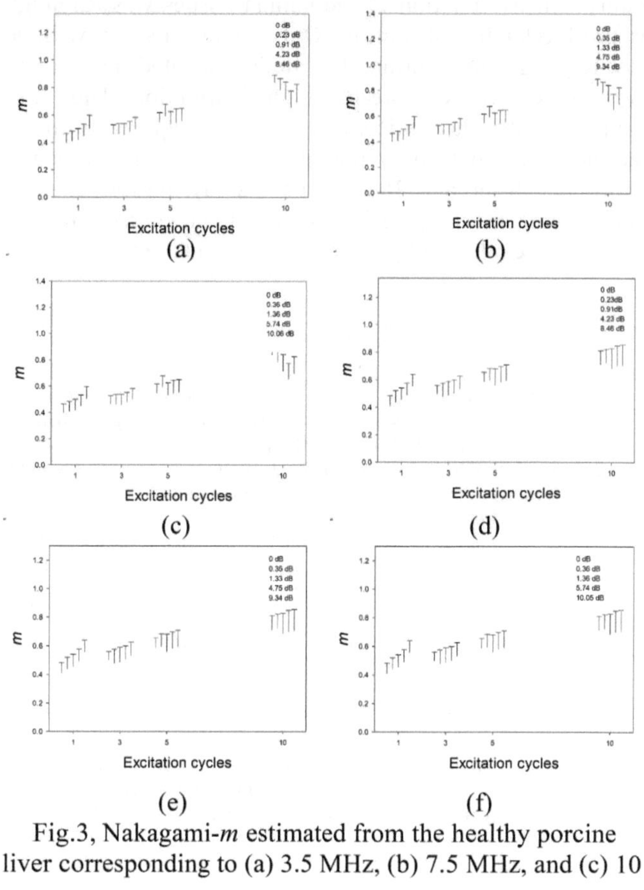

Fig.3, Nakagami-*m* estimated from the healthy porcine liver corresponding to (a) 3.5 MHz, (b) 7.5 MHz, and (c) 10

MHz transducers; those of the pathology porcine liver by using (d) 3.5 MHz, (e) 7.5 MHz, and (f) 10MHz transducer.

IV. Discussion

The effects of ultrasonic frequency, excitation cycles, and attenuation on the estimation of Nakagami-*m* were investigated by using different concentration of the phantoms and porcine livers. The Nakagami-*m* was influenced not only by the scatterers but also by the characteristic of the trasnducers. Therefore, the Nakagami-m was increased with the increased resolution cell, and it was increased with the pulse length and the lateral profile. Hence, the low frequency of the transduver is often with a wider lateral profile than the high frequency of the transducer. The results showed the pulse duration of 3 excitation cycles was better than 1, 5 and 10 cycles. The effect of the attenuation can be reduced by the increasing excited cycles and the great amount of the scatterers. However, the increased excitation cycle not only reduced the attenuation but also decreased the axial resolution of the transducer. In generally, the probability density function of the human tissues was Rayleigh or pre-Rayleigh distribution. Only in few tissues will be post-Rayleigh distribution. The study indicated the porcine liver tissues all were pre-Rayleigh distribution. The Nakagami-*m* of the ultrasonic envelopes were acquired from the pathological and healthy porcine livers were no statically significant difference. Based on the study, performance of using Nakagami imaging in practical tissue characterization was further evaluated by animal experiments *in vivo.*

References

1. R. C. Molthen, P. M. Shankar, and J. M. Reid (1995), "Characterization of ultrasonic B-scans using non-Rayleigh statistics," Ultrasound Med. Biol., vol. 21, pp. 161-70.
2. S. W. Smith, H. Lopez, and W. J. Bodine, Jr., "Frequency independent ultrasound contrast-detail analysis (1985)," Ultrasound Med. Biol., vol. 11, pp. 467-77.
3. P. M. Shankar (1995), "A model for ultrasonic scattering from tissues based on the K distribution," Phys. Med. Biol., vol. 40, pp. 1633-49.
4. P. Mohana Shankar (2000), "A general statistical model for ultrasonic backscattering from tissues," *IEEE Trans. Ultrason. Ferroelectr. Freq. Control,* vol. 47, pp. 727-736.
5. D. Nicholas, D. K. Nassiri, P. Garbutt, and C. R. Hill (1986), "Tissue characterization from ultrasound B-scan data," *Ultrasound Med. Biol.,* vol. 12, pp. 135-43.
6. P. H. Tsui and S. H. Wang, "The effect of transducer characteristics on the estimation of Nakagami parameter as a function of scatter concentration," Ultrasound Med. Biol., 30: 1345-1353, 2004.
7. P. M. Shankar, V. A. Dumane, J. M. Reid, V. Genis, F. Forsberg, C. W. Piccoli and B. B. Goldberg, "Classification of ultrasonic B-mode images of breast masses using Nakagami distribution," IEEE Trans. Ultrason. Ferroelectr. Freq. Control, 48: 569-580, 2001.
8. C. C. Huang, S. H. Wang and P. H. Tsui, "Detection of blood coagulation and clot formation using quantitative ultrasonic parameters," Ultrasound Med. Biol., 31: 1567-1573, 2005.
9. Y. H. Lin, C. C. Huang and S. H. Wang, "Quantitative assessments of burn degree by high-frequency ultrasonic backscattering and statistical model," Phys. Med. Biol., 56: 757-773, 2011.
10. P. H. Tsui and C. C. Chang, "Imaging local scatterer concentrations by the Nakagami statistical model," Ultrasound Med. Biol., vol. 33, pp. 608-19, 2007.

Microwave System Development for Medical Diagnostics

X. Zeng[1], A. Fhager[1], M. Persson[1], and H. Zirath[2]

[1] Department of Signals and Systems, Chalmers, Göteborg, Sweden
[2] Department of Microtechnology and Nanoscience, Chalmers, Göteborg, Sweden

Abstract— **This paper presents a time domain microwave system developed for medical diagnostics. An imaging test was carried out with the developed prototype and reconstructed images were compared with those obtained by using a network analyzer. Results show small difference between reconstructions, which validates the imaging functionality of the prototype.**

Keywords— **Time domain measurements, ultra-wideband, microwave imaging, medical applications, microwave receivers.**

I. INTRODUCTION

There has been an increasing interest on using microwaves for medical applications [1-4]. In comparison with other technologies, microwave method has many advantages. It is safe, comfortable, and potentially can give high resolution images.

Most of the studies carried out so far take advantage of lab instruments [2, 5], which are accurate and easy to perform. However, these instruments are bulky and expensive, and not designed with the purpose of being used at the clinic. In order to bring the technique to the point of care, a compact and efficient system is of high importance.

We aim at designing and developing a dedicated ultra-wideband microwave system for medical diagnostics. The system should be compact, fast and have sufficient accuracy. Our long term goal is to have a system on chip.

The work presented in this paper is the preliminary step towards the goal. We developed a time domain system based on off-the-shelf components and tested its imaging functionality.

II. METHOD

The main challenge in designing a time domain system is the insufficient sampling rate and bandwidth of an analog-to-digital converter (ADC) [6]. The system design is therefore based on an equivalent time sampling method in order to generate the required sampling rate. The basic principle of this sampling method can be found in [6]. In order to solve the problem of insufficient bandwidth, a wide band track-and-hold (T/H) circuit is used ahead of the ADC. It "track" the received signal and at the sample time "freezes" the signal level, holding it for a certain period of time in order to allow for the ADC to sample.

A prototype is developed and the development mainly focuses on the receiver side. Fig. 1 shows the block diagram of the developed system. An arbitrary waveform generator (AWG) is used to generate a pulse. A wide band power amplifier (PA) from Mini-circuits is used to amplify the generated pulse. The T/H-ADC assembly is used to sample and digitize the received signal.

A four channel direct digital synthesizer (DDS) is used for generating sampling clocks with the same frequency but different phases, named phase 1, phase 2 and phase3 respectively. Phase 1 and phase 2 are a differential pair and they are used to trigger the T/H. Phase 3 is the ADC sampling clock and its phase has to be well tuned in relative to the T/H trigger in order to make the system work properly. An Agilent synthesizer is to provide synchronisation between the pulse generation and the sampling.

The entire measurement procedure is controlled by the FPGA. A clock is fed from the AWG to the FPGA in order to set the initial timing of the data storage. Fig. 2 gives a picture of the system receiver which is composed of off-the-shelf components.

III. RESULTS

A simple imaging test was performed with the developed system in order to validate the system and its functionality.

A two dimensional antenna array composed of 20 monopoles was used. The twenty antennas were evenly distributed on a circle with a radius of 10 centimetres [2]. A plastic cup filled with vegetable oil, was used as the imaging target, and placed at the centre of the antenna array. The plastic cup was cylindrical-shaped, but with gradually increasing radius. The bottom diameter and top diameter of the cup were 55 mm and 75 mm respectively. In the frequency range of interest, the dielectric constant of the oil is in the range 2~3 and the conductivity is close to zero. Fig. 3 shows the imaging setup.

With the imaging system, a pulse generated by the AWG is transmitted by one of the antennas into the object under test, and the scattered field is acquired by the remaining antennas. The acquired signals are measured by means of the T/H-ADC assembly and the data is stored in the FPGA for further processing. This process is repeated until all the

H. Mindedal and M. Persson (eds.), *16th Nordic-Baltic Conference on Biomedical Engineering*,
IFMBE Proceedings 48, DOI: 10.1007/978-3-319-12967-9_32

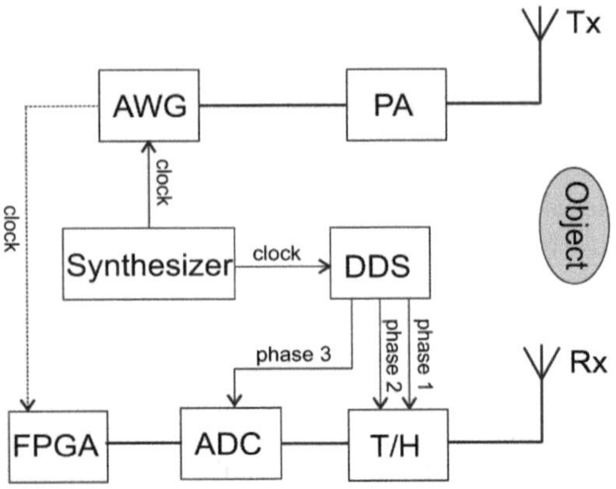

Fig. 1 Block diagram of the developed prototype

Fig. 2 Hardware of the developed prototype

antennas have been used for transmitting. A mechanical switching matrix was employed to select different transmitting and receiving antenna pairs.

Two groups of 20 × 19 data sets were obtained when the number of average in the measurements were $AVG_N = 1$ and $AVG_N = 16$ respectively. The same measurements were taken for an empty antenna system, in order to carry out the calibration of the measured data [2].

Dielectric profiles reconstructed were shown in Fig. 4. For comparison the same object was measured and reconstructed based on measurements with a vector network analyzer.

Fig. 3 Imaging setup. A plastic cup of vegetable oil is put at the center of the antenna array.

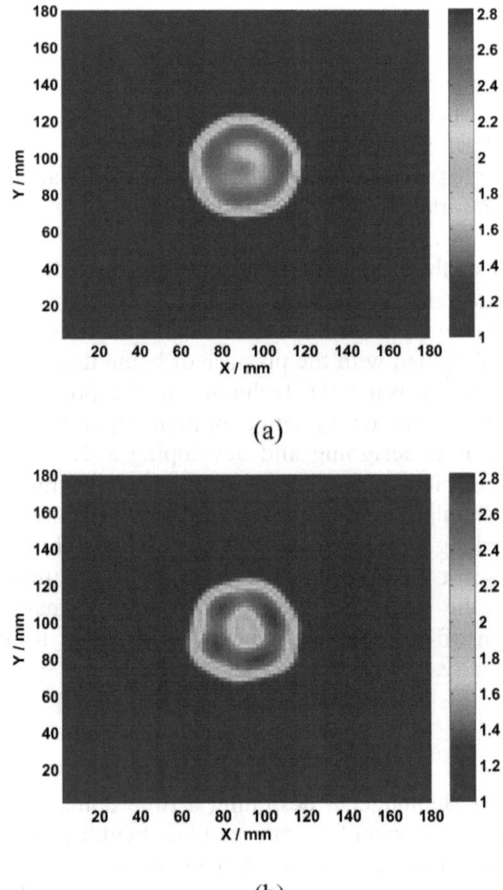

(a)

(b)

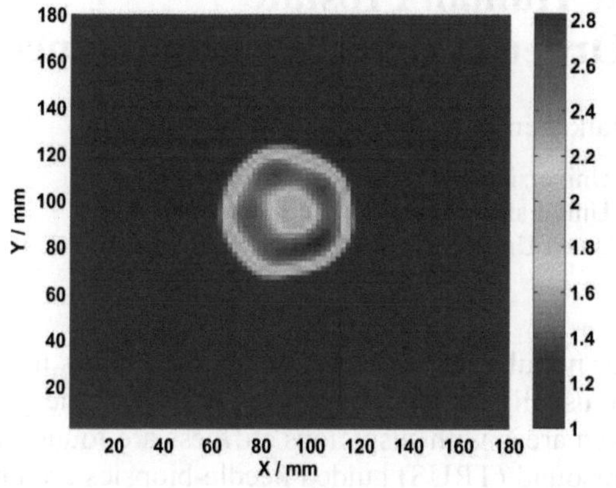

(c)

Fig. 4 Imaging results obtained by using (a) a network analyzer and (b) and (c) developed prototype when AVG_N = 1 and 16, respectively. The spectral content used in the reconstruction is centered at 3 GHz with 3 GHz bandwidth.

The imaging results obtained with the prototype and the VNA are very similar, which confirms the imaging functionality of the developed prototype.

The results also show that the averaging of measurements does not result in much improvement of the image quality, which suggests that the accuracy obtained when $AVG_N = 1$ is good enough for the imaging.

CONCLUSIONS

We have presented a preliminary prototype developed for medical applications. Its imaging functionality has been confirmed with an imaging test.

The results are in an early development phase of the system, yet it shows promising performance characteristics.

Next step, we will perform imaging tests on phantoms with more realistic features to human tissues.

The prototype presented in this work was only partly customized. A completely customized system will be developed in the future with off-the-shelf and self-designed components.

ACKNOWLDEDGMENT

This work was supported in part by the Swedish Agency for Innovation Systems within the Chalmers Antenna Systems VINN Excellence Centre and in part by the Swedish Foundation for Strategic Research within the Strategic Research Center Charmant.

REFERENCES

[1] Meaney P, Fanning M, Li D et al.(2000) A clinical prototype for active microwave imaging of the breast IEEE Trans. Microwave Theory Tech 48: 1841–1853

[2] Fhager A, Hashemzadeh P, and Persson M et al. (2006) Reconstruction quality and spectral content of an electromagnetic time-domain inversion algorithm. IEEE Trans. Biomed. Eng 53: 1594–1604

[3] Fhager A, Mckelvey T and Persson M (2010) Stroke detection using a broad band microwave antenna system. EUCAP 2010. Barcelona, Spain, 2010.

[4] Persson M, Zeng X and Fhager A (2011) Microwave imaging for medical applications. EUCAP 2011. Rome, Italy.

[5] Zeng X, Fhager A, Linner P et al. (2011) Experimental investigation of the accuracy of an ultrawideband time-domain microwave-tomographic System," IEEE Tran. Instrum. Meas 60: 3939–3949

[6] Application Note 1608. What is the Difference Between an Equivalent Time Sampling Oscilloscope and A Real-Time Oscilloscope? at http://www.home.agilent.com/

Author: Xuezhi Zeng
Institute: Chalmers University of technology
Street: Hörsalsvägen 11, 412 96
City: Gothenburg
Country: Sweden
Email: xuezhi@chalmers.se

Initial Measurements on Whole Human Prostate *ex vivo* with a Tactile Resonance Sensor in Order to Detect Prostate Cancer

Anders P. Åstrand[1,2], Britt M. Andersson[1,2], Ville Jalkanen[1,2], and Olof A. Lindahl[2,3]

[1] Department of Applied Physics and Electronics, Umeå University, SE-90187 Umeå, Sweden
[2] Center for Biomedical Engineering and Physics, Umeå University, SE-90187 Umeå, Sweden
[3] Department of Radiation Sciences, Biomedical Engineering, Umeå University, SE-90187 Umeå, Sweden

Abstract—**Prostate cancer (PCa) is the most common form of cancer among the male population in Europe and the USA. PCa can be suspected by a blood test for a specific prostate antigen, a PSA-test, followed by a digital rectal examination (DRE). The objective with the DRE is to investigate the presence of stiff nodules on the prostate. Stiff nodules can indicate PCa and biopsies are taken from the suspicious parts of the prostate using guidance of a transrectal ultrasound. Microscopic evaluation of the biopsies is used for final diagnosis. Superficial tumor growth on, and beneath the surface of the gland is of special interest as it suggests that the cancer has spread to other parts of the body.**

Tactile resonance sensors can be used to distinguish between areas of different stiffness in soft tissue. The aim was to detect tumors on, and beneath the surface of a whole human prostate *ex vivo*.

A tactile resonance sensor system (TRSS) based on a piezoelectric resonance sensor and a force sensor has been used to detect areas with increased stiffness in soft tissue. The TRSS has a rotatable sample holder for measurements on spherical shaped samples. Stiffness measurements were made on samples of porcine muscle tissue with embedded stiff silicone nodules placed under the surface. Further measurements were made on a resected whole human prostate with PCa.

The results showed that through the measured stiffness parameter, the stiff silicone nodules placed down to 4 mm under the surface could be detected. The measurements on the prostate showed that elevated values of the stiffness parameter correlated ($p < 0.05$) with areas in the anterior of the prostate where cancer tumors were detected by histopathological evaluation. The tumors were significantly stiffer than the healthy tissue in the dorsal region.

The results are promising for further development of a clinically useful instrument to detect superficial PCa.

Keywords— **Prostate cancer, Tissue stiffness, Resonance sensors.**

I. INTRODUCTION

Research in new and improved methods for early detection of prostate cancer (PCa) is very important. The most common screening methods to indicate PCa is the PSA-test and the digital rectal examination (DRE), where the physician palpates the prostate trough the rectum to detect stiff areas or nodules. It has been shown that tumors in the prostate are usually stiffer compared to healthy tissue [1, 2, 3, 4, 5]. When areas with suspicious stiffness are found, transrectal ultrasound (TRUS) guided needle-biopsies are taken and microscopically evaluated to obtain a histological diagnosis [6]. The biopsies fail to detect 10-30% of the PCa, and as the DRE is a subjective method dependent on the physician's experience, an objective method and a quantitative parameter related to the prostate tissue would be preferred [7].

In this study, a tactile resonance sensor system (TRSS), earlier described in [8, 9, 10] based on the principle of an oscillating piezoelectric element has been used for stiffness measurements. The principle of using a piezoelectric element as a resonance sensor in this type of setting has been described in the early 1990s by Omata and Terunuma [11]. In more recent studies [4, 12] the technology has been used to measure stiffness variations related to heterogeneous prostate histology with malignant tissue. However, those studies were made on slices of prostate tissue, whereas in this study the measurements were made on a whole prostate *ex vivo*.

The aim of this study was at first to evaluate the ability of the TRSS to detect stiffer volumes at a distance beneath the surface of spherically shaped samples of porcine muscle with embedded stiff silicone nodules. Finally measurements were made on one fresh prostate gland *ex vivo*, from a radical prostatectomy patient.

II. MATERIALS AND METHODS

The sensors in the TRSS consisted of a piezoelectric element made of lead zirconate titanate PZT (Type 7A, Morgan electro mechanics, Bedford, Ohio, USA) and a preloaded force sensor (PS-05KC, Kyowa, Tokyo, Japan).The PZT-element was in the shape of a 15 mm long cylinder with an outer diameter of 5 mm, and an inner diameter of 3 mm with a hemispherical tip of polyether-ether-ketone (PEEK) with the radius of 5 mm for sample contact (Fig. 1). Three motor controller translation stages for horizontal and

H. Mindedal and M. Persson (eds.), *16th Nordic-Baltic Conference on Biomedical Engineering*,
IFMBE Proceedings 48, DOI: 10.1007/978-3-319-12967-9_33

vertical movements were controlled by a LabView® program.

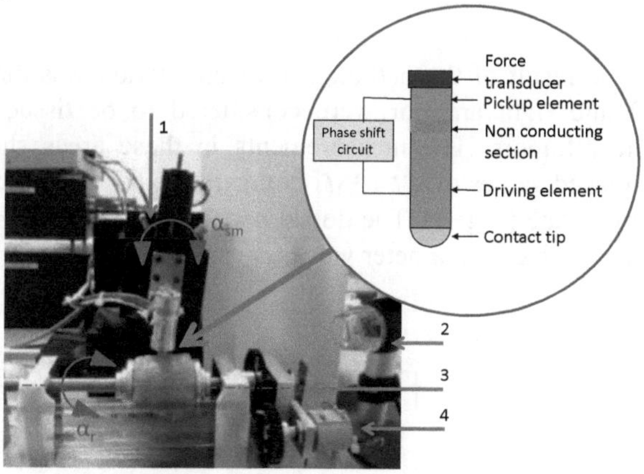

Fig. 1 The TRSS with a sample of porcine muscle mounted in the rotatable holder. The insertion shows the piezoelectric element inside the sensor head. 1) The rotational stage which control the contact angle α and the angle of the sensor movement, α_{sm}. 2) USB-microscope. 3) The cantilever strain gauge. 4) A sensor for measuring the rotation angle α_r.

The measured parameters from the TRSS were the change in resonance frequency of the PZT-element, Δf, and the applied force, F, during the indentation, I, into the measured sample under study (porcine tissue and human prostate tissue). A stiffness parameter, $|\partial F/\partial \Delta f|$, can be calculated from the measured Δf and F as a function of the indentation depth I [13].

The measurements in this study were all made with an indentation velocity υ_i = 4 mm s^{-1}, and a total indentation depth I_{tot} = 1.0 mm. The logged data for each measurement were Δf, F, and the rotation angle, α_r. From the sampling rate of 1 kHz and υ_i, the data for the I of interest could be calculated. In this report, data at I = 0.6 mm was analysed. From Δf and F, $|\partial F/\partial \Delta f|$ was calculated through linear regression for a corresponding interval $I = 0.6 \pm 0.2$ mm [8, 9].

A. Porcine tissue sample

The initial measurements were made on commercially available porcine muscle from tenderloin that was used as soft tissue phantoms to mimic prostate tissue. They were cut into spherical shapes, approximately 40 mm in diameter. Small spherical nodules of stiff silicone with Shore hardness 88 (scale 000) and the diameter 6 mm, were inserted through an incision at a distance, d, under the surface (Fig. 2A). The measurements were made with a vertical sensor movement, perpendicular to the surface. The samples were rotated ± 30° from the position of the stiff nodule in steps of $\Delta\alpha_r$ = 5°. Saline solution (Sodium Chloride 9 mg/ml) was sprayed on the surface of the samples between measurements in order to keep it moist.

B. Whole human prostate sample

The final measurements were made on one human prostate gland, obtained with written consent from a 72 years old patient undergoing a radical prostatectomy (Fig 2B). Ethical approval was given by the Ethics Committee at Umeå University (Dnr 03-423).

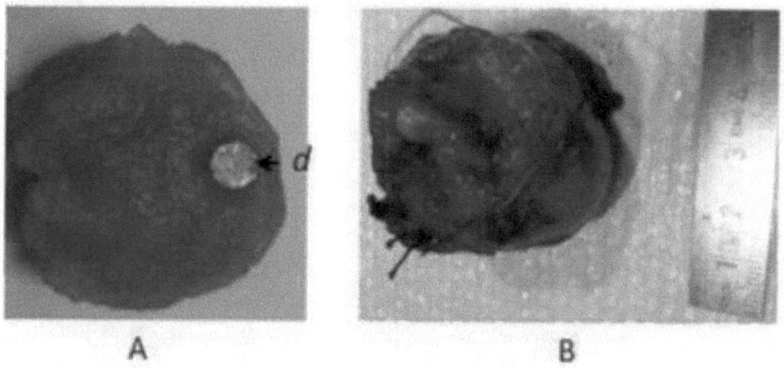

Fig. 2 A) Porcine tissue with an inserted silicone nodule at a depth d from the surface. B) The prostate gland, fresh from surgery.

For orientation purposes, the prostate was stained according to the routines for histopathological procedures before the measurements could take place. The dorsal (backside) aspect was stained yellow, the anterior (front) left side red, and the right side green. The prostate was placed in the rotatable holder with the urethra in the horizontal axial direction. During the measurements, the prostate was rotated and measurements were made at approximately every 10° (i.e. $\Delta\alpha_r$ = 10°) with the sensor movement kept vertical. The measurements were made along the great circle and covered all three regions of the prostate, resulting in 40 measurements. The prostate was kept moist by spraying it with saline solution. After the measurements, the prostate was returned to the pathologist for further examination and processing according to their standard procedures. One slice was cut along the line where the measurements were made, and this slice was dehydrated and embedded in paraffin from which a 5 µm thin slice was stained with Haematoxylin-Eosin. This slice was examined by light microscope, and areas with tumors were marked by the pathologist.

III. Results

A. Measurements on porcine tissue

The porcine muscle tissue sample contained the stiff silicone nodule embedded at $d = 3 \pm 0.5$ mm under the surface. Figure 3 shows the measured $|\partial F/\partial\Delta f|$ at I = 0.6 mm and the silicone nodule was detected, as well as another stiff area at $\alpha_r = 30°$, which was shown to be a hidden tendon.

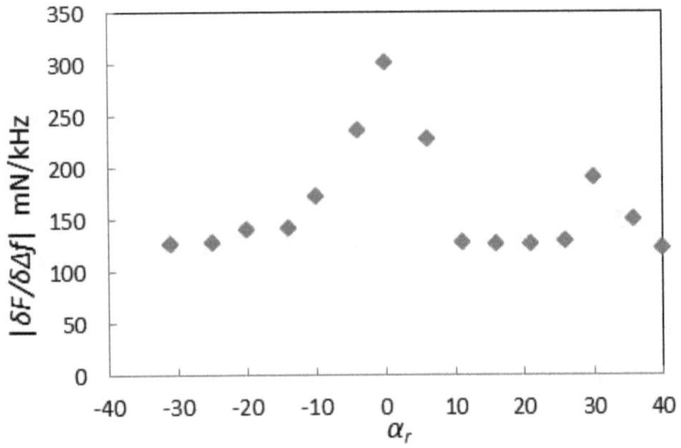

Fig. 3 Measured $|\partial F/\partial\Delta f|$ on a sample of porcine muscle tissue with a hidden nodule of stiff silicone found approximately at $\alpha_r = 0°$ and located at a depth $d = 3 \pm 0.5$ mm under the surface.

B. Measurements on whole human prostrate

The dehydrated tissue slice and a photomicrograph of the prostate are shown in Figure 4A and 4B. The pathologist had marked a large area with presence of cancer in the anterior region of the prostate. The different colors that the prostate was stained with are partly visible in the periphery.

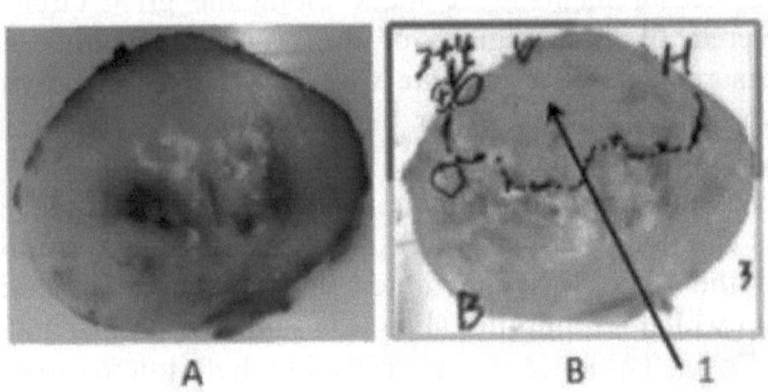

Fig.4 The prostate is shown in A and B. A) The slice of the prostate tissue that was embedded in paraffin. B) Photomicrograph of the corresponding section. The surrounding frame is colored as the stained periphery, the left anterior (red), right anterior (green) and dorsal (yellow). The arrow, 1) point at a large area in the anterior part of the prostate containing cancer tumors. In the left anterior, the pathologist has marked tumor with a Gleason score 7 (3+4).

The result of the pathologist's examination was that the left and right anterior were considered to be tissue with cancer tumors. The measurements in these areas showed increased values of $|\partial F/\partial\Delta f|$ compared to the other areas of the prostate (Fig. 5). The dorsal part of the prostate showed lower stiffness parameter values.

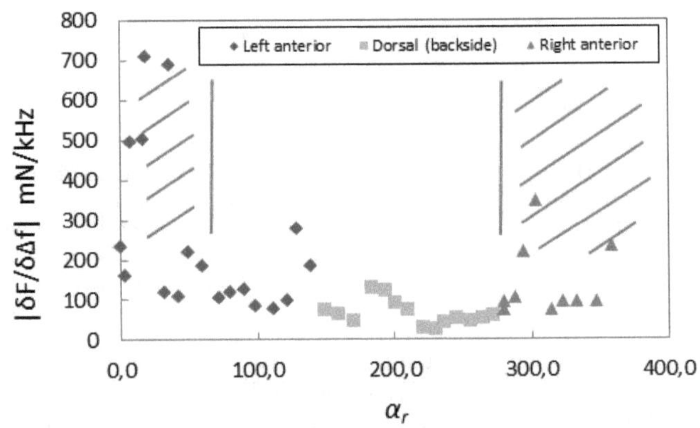

Fig. 5 The stiffness parameter from the measurements on the prostate. The marked areas in the figure show the tissue section where the pathologist marked the presence of tumor tissue. $|\partial F/\partial\Delta f|$ was measured at approximately $\Delta\alpha_r = 10°$ along a line around the prostate gland. The colors in the figure represent the different stained areas of the prostate as described above.

The results show that there is a statistical significant difference ($p < 0.05$) in stiffness values between the measurements in the area that was considered to have presence of cancer tumors (n = 17) compared to the stiffness measurements in the other parts of the prostate (n = 21).

IV. Dicussions

The surfaces on biological tissue samples are often uneven which can lead a non-perpendicular contact angle between the sensor tip and the surface of the sample. This can lead to errors in the measured parameters [8]. For this reason, photographs were taken from two directions at the time of each measurement. By studying the photographs, it was estimated that the deviation from the perpendicular was < 8° for all measurements. According to an earlier study [8], such small variations did not significantly affect the measured values of Δf and F and thus the $|\partial F/\partial \Delta f|$.

Additional measurements will be made on whole human prostates which will be complemented with histopathological analysis for presence of cancer tumors and tissue characterization. This will be published in the near future.

V. CONCLUSIONS

This initial study reports on stiffness measurements on soft tissue models with embedded stiff nodules mimicking cancer tumors, and measurements on a resected whole human prostate gland with presence of prostate cancer. The measurements were made with a tactile resonance sensor system with a rotatable holder. Stiff silicone nodules placed down to 4 mm under the surface of porcine muscle tissue could be detected with an indentation of the sensor of 0.6 mm. The measurements on the prostate showed areas with elevated stiffness parameter values ($p < 0.05$) compared to the dorsal part. This agreed with the location of the histologically determined cancer tumors.

This is promising for the development of a clinically useful instrument to detect superficial prostate cancer.

ACKNOWLEDGMENT

The authors wish to express their gratitude to Kerstin Almroth, Research Assistant at the Department of Surgical and Perioperative Sciences, Urology and Andrology at Umeå University, Wanzhong Wang, Pathologist, Pernilla Andersson, Research Assistant, Birgitta Ekblom, Research Assistant and Susanne Gidlund, Research Assistant, at the Department of Clinical Pathology at Umeå University Hospital, Morgan Nyberg at Department of Engineering Sciences and Mathematics, Luleå University of Technology.

The study was supported by grants from objective 2 North Sweden-EU Structural Fund.

REFERENCES

1. Byar DP, Mostofi FK and the Veterans Administration Cooperative Urological Research Group 1972 Carcinoma of the prostate: prognostic evaluation of certain pathological features in 208 radical prostatectomies *Cancer* **30** 5-13
2. Gwilliam JC, Pezzementi Z, Jantho E, Okamura AM, Hsiao S 2010 Human vs. robotic tactile sensing: detecting lumps in soft tissue *IEEE Haptics Symposium 2010 (Waltham, MA)* 21-28. ISBN 978-1-4244-6821-8
3. Tiwana MI, Redmond SJ, Lovell NH 2012 A review of tactile sensing technologies with applications in biomedical engineering *Sensors and Actuators* **A179** 17-31
4. Jalkanen V, Andersson BM, Bergh A, Ljungberg B and Lindahl OA 2006 Resonance sensor measurements of stiffness variations in prostate tissue in vitro – a weighted tissue proportion model *Physiol. Meas.* **27** 1373-86
5. Krouskop TA, Wheeler TM, Kallel F, Garra BS, Hall T 1998 Elastic moduli of breast and prostate tissues under compression *Ultrason Imaging* 20 260-274
6. Zhang M, Nigwekar P, Castaneda B, Hoyt K, Joseph JV, di Sant Agnese A, Messing EM, Strang JG, Rubens D and Parker KJ 2008 Quantative characterization of viscoelastic properties of human prostate correlated with histology *Ultrasound in med & Biol* vol **34** no 7 1033-1042
7. Ahn B-M, Kim J, Ian L, Rha K-H and Kim H-J 2010 Mechanical property characterization of prostate cancer using a minimally motorized indenter in an ex vivo indentation experiment *Urology* vol **76** no 4 1007-1011
8. Åstrand AP, Jalkanen V, Andersson BM, Lindahl OA 2013 Contact angle and indentation velocity dependency for a resonance sensor – evaluation on soft tissue silicone models *J Med Eng Technol* **37(3)** 185-196
9. Åstrand AP, Jalkanen V, Andersson BM and Lindahl OA 2014 Detection of stiff nodules embedded in soft tissue phantoms, mimicking cancer tumours, using a tactile resonance sensor. *Journal of Biomedical Science and Engineering* **7** 181-193
10. Åstrand AP 2012 *A Flexible Resonance Sensor System for Detection of Cancer Tissue – Evaluation on Silicone*. Lic. Thesis, Chapter 4, 5-8. Umeå University, Sweden. ISBN 978-91-7459-477-5
11. Omata S, Terunuma Y 1992 New tactile sensor like the human hand and its applications *Sens Actuat.* **35** 9-15
12. Jalkanen V, Andersson BM, Bergh A, Ljungberg B and Lindahl OA 2006 Prostate tissue stiffness as measured with a resonance sensor system: a study on silicone and human prostate tissue in vitro *Med. Biol. Eng. Comput.* **44** 593-603
13. Jalkanen V, Andersson BM, Bergh A, Ljungberg B and Lindahl OA 2008 Explanatory models for a tactile resonance sensor system – elastic and density-related variations of prostate tissue in vitro *Physiol. Meas.* **29** 729-45

Author: Anders Åstrand
Institute: Department of Applied Physics and Electronics
City: Umeå University
Country: Sweden
Email: anders.astrand@tfe.umu.se

Author Index

T

U

W

Z

Keyword Index

N

O

P

Q

S

T

U

W

MIX
Papier aus verantwortungsvollen Quellen
Paper from responsible sources
FSC® C105338

If you have any concerns about our products,
you can contact us on
ProductSafety@springernature.com

In case Publisher is established outside the EU,
the EU authorized representative is:
Springer Nature Customer Service Center GmbH
Europaplatz 3, 69115 Heidelberg, Germany

Printed by Libri Plureos GmbH
in Hamburg, Germany